Grant of a Patent, Opposition, Appeal, and Invalidity

Thomas Heinz Meitinger

Grant of a Patent, Opposition, Appeal, and Invalidity

The Proceedings before the DPMA, the EPO, and the Federal Patent Court (BPatG)

Springer

Thomas Heinz Meitinger
Meitinger Patentanwalts GmbH
Munich, Germany

ISBN 978-3-662-72555-9 ISBN 978-3-662-72556-6 (eBook)
https://doi.org/10.1007/978-3-662-72556-6

Translation from the German language edition: "Patenterteilung, Einspruch, Beschwerde und Nichtigkeit." by Thomas Heinz Meitinger, © Der/die Herausgeber bzw. der/die Autor(en), exklusiv lizenziert an Springer-Verlag GmbH, DE, ein Teil von Springer Nature 2025. Published by Springer Berlin Heidelberg . All Rights Reserved.

This book is a translation of the original German edition "Patenterteilung, Einspruch, Beschwerde und Nichtigkeit." by Thomas Heinz Meitinger, published by Springer-Verlag GmbH, DE in 2025. The translation was done with the help of an artificial intelligence machine translation tool. A subsequent human revision was done primarily in terms of content, so that the book will read stylistically differently from a conventional translation. Springer Nature works continuously to further the development of tools for the production of books and on the related technologies to support the authors.

This Springer imprint is published by the registered company Springer-Verlag GmbH, DE, part of Springer Nature.
The registered company address is: Heidelberger Platz 3, 14197 Berlin, Germany

Preface

The patent granting procedure before the Patent Office is primarily intended to determine the broadest possible scope of protection for the applicant in light of the prior art. During the granting procedure, it is usually necessary to respond to several official communications and to regularly amend the original claims in order to obtain a set of claims that is eligible for grant. This book describes the official granting procedure and provides guidance to ensure that a patent is granted promptly and successfully. If the patent cannot be granted, the decision of the Patent Office to reject the application can be challenged before the Federal Patent Court by means of an "applicant's appeal."

After a patent has been granted, it can be challenged by any third party in contentious proceedings before a patent office or a court. Opposition proceedings against German patents take place before the German Patent and Trade Mark Office. Oppositions and appeals against oppositions concerning European patents are heard before the European Patent Office. Appeals against oppositions and nullity proceedings against German patents or the German part of a European patent are heard before the Federal Patent Court. Nullity appeal proceedings against German patents are conducted before the Federal Court of Justice in Karlsruhe. In addition, there are official cancellation proceedings for utility models, trademarks, and design rights. All these proceedings before the DPMA (German Patent and Trade Mark Office), the EPO (European Patent Office), the BPatG (Federal Patent Court), the BGH (Federal Court of Justice), the EUIPO (European Union Intellectual Property Office), and the UPC (Unified Patent Court) are described in this book.

The first chapter provides an overview of industrial property protection and its procedures. Chapters 2 to 15 describe the procedures based on technical intellectual property rights, i.e., patents and utility models. Chapter 16 deals with trademarks, and Chap. 17 with design rights. The main focus of the book is therefore on procedures relating to technical intellectual property rights.

Munich
in March 2025

Dr. Thomas
Heinz Meitinger

Contents

About the Author

Dr. Thomas Heinz Meitinger is a German and European patent attorney. He is the managing director of Meitinger Patentanwalts GmbH. Meitinger Patentanwalts GmbH is a medium-sized patent law firm based in Munich. After studying electrical engineering in Karlsruhe, he initially worked as a development engineer. Subsequent positions included roles as production manager and technical director in medium-sized companies. Dr. Meitinger regularly publishes scientific articles and writes technical books on intellectual property law and technical subjects. He gives lectures on patent, trademark, and design law. Dr. Meitinger holds the degrees Dipl.-Ing. (Univ.) and Dipl.-Wirtsch.-Ing. (FH). In addition, he holds the following master's degrees: LL.M., LL.M., MBA, MBA, M.A., and M.Sc.

Abbreviations

BGH	Federal Court of Justice (Bundesgerichtshof)
BPatG	Federal Patent Court (Bundespatentgericht)
DPMA/GPTO	German Patent and Trademark Office (Deutsches Patent- und Markenamt)
EPA/EPO	European Patent Office (Europäisches Patentamt)
EPG/UPC	Unified Patent Court (Einheitliches Patentgericht)
EPÜ/EPC	European Patent Convention (Europäisches Patentübereinkommen)
EUIPO	European Union Intellectual Property Office

Fundamentals of Industrial Property Protection 1

Contents

T. H. Meitinger, *Grant of a Patent, Opposition, Appeal, and Invalidity*,
https://doi.org/10.1007/978-3-662-72556-6_1

The introduction chapter presents the main ex parte (only the applicant is involved) and inter partes (applicant versus opposing party) proceedings, as well as the institutions before which these proceedings are conducted. In addition, the most important grounds for revocation, which are negotiated in these proceedings, are listed.

The requirements for patentability are described, and the different types of technical intellectual property rights—patent application, patent, and utility model—are introduced. Both German and European patent law recognize the official patent granting procedure, in which the relevant patent office subjects the application to an official examination for protectability. At the end of the examination procedure, either a patent is granted or the patent application is rejected.

Utility model law does not provide for an official examination procedure. The legal validity of a utility model is only examined in a cancellation procedure before the DPMA or in a utility model appeal procedure, as a legal remedy of the cancellation procedure, before the Federal Patent Court.

1.1 Industrial Property Protection

Industrial property protection deals with the intellectual property rights that inventors and companies register with a patent office to legally safeguard their business activities. The best-known intellectual property rights are patents, utility models, and trademarks. Design rights are of lesser importance.

1.2 Patent Application

A patent application is assigned a filing date if it meets the formal requirements of the patent office.[1] A patent application is an unexamined intellectual property right and does not confer any right to prohibit. Anyone is permitted to use the inventions described in patent applications. The holder may only claim compensation for use from the date of publication of the patent application.[2] The claim for compensation does not constitute damages and is lower.

[1] Section 35(1) Patent Act.

[2] Section 33(1) Patent Act.

1.3 Patent

Within a period of seven years from the filing date, a request for examination must be filed.[3] Otherwise, the application lapses. The request for examination initiates the granting procedure, at the end of which a patent may be granted. The patent grants the patent holder a right to injunctive relief against any unauthorized use of the patented invention.[4]

The right to injunctive relief exists in the case of an imminent first infringement, for example if an unauthorized party has already announced that they will not respect the patent, and in the case of a risk of recurrence. A risk of recurrence exists if a first patent infringement has already occurred.

However, the patent holder's right to injunctive relief has its limits; in particular, private use cannot be prohibited.[5] Furthermore, experiments with the technical teaching of a patent are permitted in order to better understand the patented invention.[6]

1.4 Utility Model

In utility model law, there is no official examination procedure regarding the legal validity of the described invention. Only a request for an official search of the prior art can be filed, which is used to assess legal validity.[7]

The patent office examines the utility model application for formal deficiencies before entry in the register, in particular whether the applicant is named, a request for entry of the utility model in the register has been filed, and a description of the invention is provided.[8] If these formal requirements are met, the utility model is entered in the register of the patent office.[9]

Although the utility model is an unexamined intellectual property right, the holder is nevertheless granted a right to prohibit analogous to a patent.[10] However, before asserting a claim for injunctive relief, it is advisable to examine the legal validity of the relevant utility model. In particular, an expert opinion on the novelty and inventive step of the utility model should be prepared.

---

[3] Section 44(2) sentence 1 Patent Act.

[4] Section 9 sentence 2 Patent Act.

[5] Section 11 no. 1 Patent Act.

[6] Section 11 no. 2 Patent Act.

[7] Section 7(1) Utility Model Act.

[8] Section 4(3) nos. 1, 2, and 4 Patent Act.

[9] Section 8(1) sentence 1 Patent Act.

[10] Section 11(1) sentence 2 Patent Act.

1.5 Trademark

A trademark right valid throughout Germany can be obtained by entry in the register of the German Patent and Trademark Office. An EU-wide trademark can be applied for at the EUIPO.

1.6 Design Law

Design protection can be obtained for a two- or three-dimensional appearance of a product, whereby the special lines, contours, colors, shape, or surface structure are protected.[11]

1.7 Patent Grant Procedure

The patent grant procedure (Synonym: Patent Prosecution or simply Prosecution; legal disputes are referred to as litigation) serves to examine an invention in a patent application for patentability. The technical teaching of a patent application must, in particular, be novel and based on an inventive step.[12] The aim of a patent grant procedure is to find a claim formulation that meets the requirements of patent law in light of the prior art.

A patent grant procedure is conducted in writing, in which the applicant must respond to office actions from the patent office by means of "responses to office actions." The office actions list the deficiencies that prevent the granting of a patent.[13] The applicant is given a deadline to respond, usually four months. The deficiencies in the patent application described by the office must be remedied. Otherwise, the patent application will be rejected.[14]

Upon request, the applicant will be summoned to an oral hearing at the end of the grant procedure if rejection of the application is imminent. This is then the last opportunity to obtain a patent and should be taken seriously. The request must be made in writing. It is advantageous to include the request in the response to the office action, for example, with the following wording:

> *"However, if the new patent request is not considered allowable and rejection of the application is contemplated, an oral hearing is requested as an auxiliary measure."* (Request for oral hearing pursuant to Section 46(1) sentence 3 Patent Act).

[11] Section 1 no. 1 Design Act.

[12] Section 1(1) Patent Act.

[13] Section 45(1) sentence 1 Patent Act.

[14] Section 48 sentence 1 Patent Act.

1.8 Opposition Procedure

The opposition procedure serves the official review of a granted patent in contentious proceedings. A substantiated opposition filed within the nine-month opposition period after publication of the grant of the patent leads to the revocation of the patent grant.[15] The opponent may, in particular, assert lack of novelty and absence of inventive step as grounds for opposition.[16]

1.9 Nullity Proceedings

A nullity proceeding serves the judicial review of a granted patent and is initiated by filing a lawsuit.[17]

1.10 Infringement Proceedings

A patent infringement can be pursued through infringement proceedings before a regional court.[18] For this purpose, special patent litigation chambers have been established in each federal state, which are exclusively responsible for patent disputes.[19]

1.11 Appeal Proceedings

A decision of the German Patent and Trademark Office can be challenged by filing an appeal before the Patent Court.[20] The decisions of the examining divisions and opposition divisions of the European Patent Office can be contested by means of an appeal.[21]

---

[15] Section 59(1) sentence 1 Patent Act.
[16] Section 21(1) no. 1 Patent Act.
[17] Section 81(1) sentence 1 Patent Act.
[18] Section 139 paragraph 1 sentence 1 Patent Act.
[19] Section 143 paragraphs 1 and 2 Patent Act.
[20] Section 73 paragraph 1 Patent Act.
[21] Article 106 paragraph 1 sentence 1 European Patent Convention (EPC).

1.12 Grounds for Opposition and Revocation

A patent will be revoked if the described invention is not new,[22] is not based on an inventive step[23] or is not industrially applicable.[24]

Furthermore, a patent can be revoked if the invention is not disclosed clearly and completely enough for a person skilled in the art to carry it out.[25] Unlawful removal (vindication) can be asserted by the entitled party.[26] Unlawful removal occurs when an invention is not filed for patenting by the inventor or their legal successor, but by an unauthorized person.

An inadmissible extension exists if the technical teaching of the patent goes beyond what was originally filed with the patent office as the application.[27]

1.13 Person Skilled in the Art

The "person skilled in the art" in patent law is an indeterminate legal concept. The person skilled in the art in patent law must always be defined when a technical matter is to be clarified. In particular, the legal fiction of the person skilled in the art is required when assessing documents of the state of the art, in questions of the feasibility of an invention (section 34 para. 4 PatG), in examining novelty (section 3 PatG), in evaluating inventive step (section 4 PatG), in interpreting patent claims to determine the scope of protection (section 14 PatG), and in assessing whether a patent infringement has occurred.[28]

The person skilled in the art in patent law is regarded as an expert in their technical field.[29] The fictional person skilled in the art in patent law possesses technical knowledge and technical skills. The technical knowledge is acquired by the person skilled in the art through standard education, in particular an engineering degree, and practical experience gained through professional activity.[30] Technical skill refers to the ability of the person

[22] Section 21 paragraph 1 no. 1 in conjunction with section 3 paragraph 1 sentence 1 Patent Act.

[23] Section 21 paragraph 1 no. 1 in conjunction with section 4 sentence 1 Patent Act.

[24] Section 21 paragraph 1 no. 1 in conjunction with section 5 Patent Act.

[25] Section 21 paragraph 1 no. 2 Patent Act.

[26] Section 21 paragraph 1 no. 3 Patent Act.

[27] Section 21 paragraph 1 no. 4 Patent Act.

[28] Dreiss: The average person skilled in the art as a standard for sufficient disclosure, patentability, and patent interpretation (GRUR 1994, 781) 782.

[29] Benkard PatG/Asendorf/Schmidt/Tochtermann, 12th ed. 2023, PatG section 4 para. 64; BGH GRUR 1959, 532 (536 f.)—Electromagnetic Stirring Device; BGH Liedl 1974/1977, 69 (78)—Target; BGH BeckRS 2009, 23384 para. 31 = BeckRS 2009, 23384.

[30] BeckOK PatR/Fitzner/Metzger, 29th ed. 15.4.2023, PatG section 3 para. 8.

skilled in the art to find a solution to a technical problem through creative considerations. However, the person skilled in the art in patent law does not possess sufficient technical skill to be considered inventive within the meaning of section 4 sentence 1 Patent Act.

The person skilled in the art in patent law is a fictional auxiliary figure,[31] which serves as a central standard of assessment in patent law.[32] However, the applicable person skilled in the art may also be a specific natural person.[33] The person skilled in the art in patent law is an average expert[34] in the technical field of the invention to be assessed and has standard education and technical skills acquired through professional activity.[35] The person skilled in the art therefore possesses average technical skills and knowledge.[36]

The determination of the person skilled in the art must not be based on the invention to be assessed, as this would constitute a retrospective evaluation of inventive step. A retrospective evaluation is not permissible, as many things appear simple in hindsight. Rather, the person skilled in the art must be determined according to the technical field to which the invention belongs.[37] The person skilled in the art is therefore not an expert for understanding the invention, but a person skilled in the relevant technical field.[38]

The person skilled in the art in patent law may be part of a team of experts if the invention concerns a subject matter that involves several different technical fields. In this case, the person skilled in the art may seek technical advice from the members of the team.[39]

---

[31] BGH GRUR 2004, 1023—Device for Separating Items from the Bottom; GRUR 2006, 663 = BlPMZ 2006, 320—Prepaid Telephone Calls.

[32] Osterrieth: The person skilled in the art in patent law GRUR 2021, 310.

[33] BGH GRUR 2004, 1023—Device for Separating Items from the Bottom; Kulhavy Mitt 2011, 179; Klett GRUR 2001, 549 (552); BGH GRUR 2018, 390—Thermal Energy Management; GRUR 2010, 602—Joint Arrangement; Meier-Beck Mitt 2005, 529; Niedlich FS König, 2003, 399; Osterrieth: The person skilled in the art in patent law (GRUR 2021, 310) 312–313.

[34] BGH GRUR 1995, 330—Electrical Plug Connection; GRUR 2004, 1023—Device for Separating Items from the Bottom; GRUR 2006, 663—Prepaid Telephone Calls; BeckRS 2005, 12061—Folding Door Unit; GRUR 2018, 390—Thermal Energy Management.

[35] BeckOK PatR/Fitzner/Metzger, 29th ed. 15.4.2023, PatG section 3 para. 8.

[36] BGH BlPMZ 1991, 159—Adhesive Bandage; GRUR 2009, 929—Abrasive Grain; GRUR 2004, 272—Diabehältnis.

[37] Benkard PatG/Asendorf/Schmidt/Tochtermann, 12th ed. 2023, PatG section 4 para. 64; BGH GRUR 1959, 532 (536 f.)—Electromagnetic Stirring Device; BGH Liedl 1974/1977, 69 (78)—Target; BGH BeckRS 2009, 23384 para. 31 = BeckRS 2009, 23384.

[38] BGH GRUR 2018, 390—Thermal Energy Management; GRUR 2010, 602—Joint Arrangement; GRUR 2016, 921—Pemetrexed.

[39] BGH GRUR 2007, 404 or BGHZ 170, 215—Carvediol II; BGH GRUR 2010, 123—Escitalopram; EPO Case T 460/87—Soluble Polyglycols/Viscosud; EPat 99/89—Circuit Arrangement for Optical Display of State Variables/Robert Boschmann GmbH.

The expertise of the notional person skilled in the art under patent law encompasses general knowledge and the customary knowledge relating to the technical field of the invention.[40] In addition, the expertise also includes the knowledge that the skilled person has acquired through their education or studies. The skilled person has also gained expert knowledge through their professional activity in the technical field. The skilled person therefore possesses both acquired knowledge and experiential knowledge.[41] The acquired knowledge is derived in particular from textbooks, reference works, and technical journals.[42]

The skilled person reads a technical teaching with their expertise and thereby recognizes certain aspects that a non-expert would not understand. The skilled person is assumed to "read along" with, for example, "means that are notoriously interchangeable in the art." For instance, the skilled person recognizes that a magnetic field can be generated by electric coils.[43] Furthermore, the skilled person is expected to be capable of carrying out standard experiments that lead to an understanding of a technical teaching.[44]

The skilled person is attributed with fundamental knowledge from related technical fields.[45] However, the skilled person does not possess detailed knowledge in the related technical fields.[46]

The expertise is not only the present knowledge of an average skilled person, but the technical knowledge available worldwide in the technical field of the invention.[47]

Through their technical skills, the skilled person is capable of creative achievement. The skilled person can, through their own considerations and limited experimentation, develop a technical subject matter. However, the skilled person is not capable of inventive activity as defined by the Patent Act.[48]

---

[40] Benkard PatG/Asendorf/Schmidt/Tochtermann, 12th ed. 2023, PatG § 4 para. 69.

[41] Ann, § 17. Novelty Ann, Patent Law, 8th edition 2022, para. 29.

[42] BGH GRUR 1998, 895 (897)—Regenbecken; BPatGE 34, 264.

[43] Dreiss: The notional skilled person as a standard for sufficient disclosure, patentability, and patent interpretation (GRUR 1994, 781) 784–785.

[44] Benkard PatG/Asendorf/Schmidt/Tochtermann, 12th ed. 2023, PatG § 4 para. 72; BGH GRUR-RS 2019, 39560 para. 121; BGH BeckRS 2016, 15771 para. 40; GRUR 2006, 666 para. 56—Stretchfolienhaube; BGH BlPMZ 1966, 234 (235)—Abtastverfahren.

[45] BGH GRUR 1997, 272 (273)—Schwenkhebelverschluss; BGH BeckRS 2008, 2725 para. 23.

[46] EPA OJ 1997, 24; Benkard PatG/Asendorf/Schmidt/Tochtermann, 12th ed. 2023, PatG § 4 para. 71.

[47] BeckOK PatR/Einsele, 29th ed. 15.7.2023, PatG § 4 para. 36.

[48] Dreiss: The notional skilled person as a standard for sufficient disclosure, patentability, and patent interpretation (GRUR 1994, 781) 783.

1.14 Novelty

Patent law applies an absolute concept of novelty.[49] Therefore, the entire state of the art that was available to the public before the filing or priority date must be taken into account when assessing novelty and inventive step.[50] The state of the art includes all publications and public uses.[51] The individual knowledge of the inventor is not relevant when assessing novelty and inventive step.

Patent law novelty is therefore not a subjective criterion.[52]

The novelty examination is conducted as a one-to-one comparison, in which the invention is compared individually and directly with each document of the state of the art. If all features of a claim can be derived from a document, that document is prejudicial to the novelty of the claim. The invention and the state of the art are to be understood from the perspective of the skilled person.[53]

A document of the state of the art is understood by the skilled person with their expertise, so that the skilled person may derive more, or even something slightly different, from a document than is literally stated therein.[54] Furthermore, the skilled person will subject the technical teaching of a document to an analysis with reasonable effort.[55] However, one should not over-interpret a document.[56]

Lack of novelty exists if all features of a claim can be derived from a single document of the state of the art.[57] If a feature necessarily results from the post-processing of the invention, it is considered to be disclosed by the state of the art.[58]

---

[49] Osterrieth, Part 4. Subject Matter, Requirements and Effect of Patent Protection, Osterrieth, Patent Law, 6th edition 2021, para. 471.

[50] § 3 paragraph 1 Patent Act.

[51] Osterrieth, Part 4. Subject Matter, Requirements and Effect of Patent Protection, Osterrieth, Patent Law, 6th edition 2021, para. 464.

[52] Osterrieth, Part 4. Subject Matter, Requirements and Effect of Patent Protection, Osterrieth, Patent Law, 6th edition 2021, para. 470.

[53] BGH GRUR 1995, 330—Elektrische Steckverbindung.

[54] BGH GRUR 1985, 330 (332)—elektrische Steckverbindung; BGH GRUR 2009, 382—Olanzapin; BGH GRUR 2013, 809—Verschlüsselungsverfahren.

[55] BGHZ 76, 97 (103)—Terephthalsäure; EPA GRUR Int. 1993, 698—"Public Accessibility".

[56] Osterrieth, Part 4. Subject Matter, Requirements and Effect of Patent Protection, Osterrieth, Patent Law, 6th edition 2021, para. 477; EPA T167/84, OJ 1987, 369—Einspritzventil.

[57] Osterrieth, Part 4. Subject Matter, Requirements and Effect of Patent Protection, Osterrieth, Patent Law, 6th edition 2021, para. 482.

[58] BGH 17.1.1980, BGHZ 76, 97 (104 ff.)—Terephthalsäure; EPA 12.2.1998, OJ 1998, 489 para. 11.1—Erythro Compounds/Novartis.

Third-party applications that were filed with the patent office before the filing or priority date of the application or patent under examination, but were only published after that date, are considered part of the state of the art and must be taken into account in the novelty examination.[59] This prevents double patenting.

1.15 Inventive Step

The assessment of inventive step is significantly more difficult compared to the examination of novelty. In the examination of inventive step, it is not only necessary to determine whether there are documents which, when considered as a whole, disclose the subject matter of the patent application. Additionally, it must be evaluated whether the skilled person would consider the technical teachings of the relevant documents to be combinable. It is therefore sometimes difficult to predict how an examiner at the Patent Office, a judge at the Federal Patent Court, or an examiner at the European Patent Office will assess the inventive step of a claim.[60]

The person skilled in the art in patent law is typically the individual who would be entrusted with development activities.[61] However, the skilled person is not a scientist,[62] but rather a practice-oriented technician.[63] The person skilled in the art in patent law is not capable of inventive activity within the meaning of § 4 sentence 1 of the Patent Act, so that all devices and methods that the skilled person can develop from the prior art are by definition not inventive under patent law. Routine developments of the prior art by the skilled person in patent law therefore cannot establish inventive step.[64]

All devices and methods that go beyond the knowledge and abilities of the skilled person based on the prior art are based on an inventive step.[65] The decision as to what is

[59] § 3 paragraph 2 Patent Act; Ann, § 17. Novelty Ann, Patent Law, 8th edition 2022, para. 4.

[60] Dreiss: The average skilled person as a standard for sufficient disclosure, patentability, and patent interpretation (GRUR 1994, 781) 785.

[61] Osterrieth GRUR 2021, 310 (311); BGH GRUR 1962, 290—Brieftauben-Reisekabine II; GRUR 1965, 138—Polymerisationsbeschleuniger; GRUR 1965, 473—Dauerwellen I; Mitt. 2003, 116—Rührwerk; BGH 8.1.2008—X ZR 110/04—Energiezuführungskette; BGH BeckRS 2008, 02725—Schleifwerkzeug für Dentalzwecke; BGH BeckRS 2009, 23384—Widerstandsschweißvorrichtung; GRUR 2009, 1039—Fischbissanzeiger; GRUR 2005, 1023—Einkaufswagen II; GRUR 2011, 1109—Reifenabdichtmittel; GRUR 2010, 513—Hubgliedertor II.

[62] RG Mitt. 1933, 156—Spannungseisen; BlPMZ 1934, 32—Phönix-Stähle II; GRUR 1942, 544—Elektronenröhre; BGH GRUR 1978, 98—Schaltungsanordnung; BlPMZ 1991, 159—Haftverband.

[63] BeckOK PatR/Einsele, 29th ed. 15.7.2023, PatG § 4 para. 36.

[64] Benkard PatG/Asendorf/Schmidt/Tochtermann, 12th ed. 2023, PatG § 4 para. 63.

[65] Klett: The averagely attentive consumer and the averagely well-trained skilled person, GRUR 2001, 549.

patentable and what is too close to the prior art must therefore be made from the perspective of the skilled person.[66]

In the examination of inventive step, the skilled person will first determine the document representing the closest prior art, and, starting from this document, will use further documents from the prior art to arrive at the invention.[67]

However, it should be noted that merely because a combination of several documents already discloses the features of an invention, this invention does not necessarily have to be obvious. Rather, the skilled person must also have had indications or suggestions to combine precisely these documents.[68]

This additional consideration is referred to at the European Patent Office as the "could-would test" and is intended to prevent an invention from appearing obvious in hindsight, even though at the time of its creation it required inventive activity.[69]

The skilled person is the central figure in the question of inventive step, as their perspective determines how a claim is to be interpreted, which documents the skilled person would consult, and what they would derive from them. Furthermore, the skilled person decides which documents are combinable and thus which technical teaching results.[70]

1.16 Industrial Applicability

The criterion of industrial applicability is of secondary importance, as only a device for which no industrial application is conceivable at all fails to meet this patentability requirement.

1.17 Impermissible Extension

An impermissible extension occurs if, as a result of subsequent amendments to the application, the content of the patent application or the patent extends beyond the subject matter originally filed with the Patent Office. This can arise, for example, when the set of claims is reformulated.[71]

---

[66] Klett: The averagely attentive consumer and the averagely well-trained skilled person (GRUR 2001, 549) 552.

[67] BGH GRUR 2018, 716—Kinderbett; GRUR 2009, 746 (748)—Betrieb einer Sicherheitseinrichtung; GRUR 2010, 407 (409)—einteilige Öse.

[68] BGH GRUR 2009, 746—Betrieb einer Sicherheitseinrichtung; GRUR 2010, 407—Einteilige Öse.

[69] Benkard PatG/Asendorf/Schmidt/Tochtermann, 12th ed. 2023, PatG § 4 para. 86.

[70] Benkard PatG/Asendorf/Schmidt/Tochtermann, 12th ed. 2023, PatG § 4 para. 34.

[71] § 21 paragraph 1 no. 4 Patent Act.

1.18 Enablement

The invention must be described in the patent application in such a way that it can be carried out by the skilled person.[72] The skilled person is expected to conduct their own experiments to a limited extent in order to achieve enablement.

1.19 Permitted Uses of a Patent

There are uses of a patent that cannot be prohibited by a patent or a legally valid utility model. In particular, private use cannot be prevented by a patent.[73] Experiments to investigate a patented invention also cannot be prohibited by a patent.

1.20 Third-Party Observations

In "third-party observations," documents from the prior art are submitted to the Patent Office that call into question the patentability of a claimed right.[74] In this way, the third party seeks to prevent the grant of the patent. The third party does not become a party to the grant proceedings.

1.21 German Patent and Trade Mark Office (DPMA)

The DPMA was founded on July 1, 1877 in Berlin as the "Imperial Patent Office." After the end of the monarchy, it was renamed the "Reichspatentamt," and after the Second World War, the Patent Office continued in Munich as the "German Patent Office." In 1998, the office received its current name, the "German Patent and Trade Mark Office."[75]

The DPMA (Deutsches Patent- und Markenamt) is responsible in Germany for the granting or registration of patents, trademarks, utility models, and design rights. In addition to its headquarters in Munich, the DPMA has offices in Jena and Berlin.[76]

[72] § 34 paragraph 4 Patent Act.

[73] § 11 no. 1 Patent Act.

[74] § 43 paragraph 3 sentence 2 Patent Act or Rule 114 paragraph 1 EPC.

[75] Wikipedia, https://de.wikipedia.org/wiki/Deutsches_Patent-_und_Markenamt, accessed on 26.2.2025.

[76] DPMA, https://www.dpma.de/dpma/wir_ueber_uns/index.html, accessed on 26.2.2025.

1.22 European Patent Office (EPO)

The European Patent Office, along with an Administrative Council, is part of the European Patent Organisation.[77] The Administrative Council is the supervisory body of the European Patent Office. The Administrative Council is composed of representatives from the member states of the European Patent Convention. Usually, the representatives are the directors of the patent offices of the respective member states.[78] The European Patent Office is the executive body of the European Patent Organisation and enables a unified patent granting procedure for the member states of the European Patent Convention (EPC).

1.23 Federal Patent Court (BPatG)

The Federal Patent Court is a federal court based in Munich.[79] The Federal Patent Court is responsible for infringement proceedings, appeals, and nullity proceedings whose subject matter concerns industrial property rights, in particular patents, trademarks, and design rights. The Federal Patent Court is the judicial authority that decides whether a patent, trademark, utility model, topography, design, or plant variety right is to be granted, or whether the protection right is to be deleted or revoked.[80]

The Federal Patent Court currently consists of 23 panels, including one Legal Appeal Panel, 7 Nullity Panels, 8 Technical Appeal Panels, 4 Trademark Appeal Panels, one Trademark and Design Appeal Panel, one Utility Model Appeal Panel, and one Appeal Panel for Plant Variety Protection Cases.[81]

The Nullity Panels deal with actions for the nullity of patents, and the Technical Appeal Panels hear appeals against decisions of the Patent Office regarding patents or patent applications. For example, appeals against the rejection of patent applications or appeals against decisions on oppositions by the German Patent Office are heard before the Technical Appeal Panels.

A special feature of the Federal Patent Court is the presence of technical judges, who do not have a legal education but possess a technical or scientific background. The technical judges participate in all proceedings before the Federal Patent Court that concern

[77] Wikipedia, https://de.wikipedia.org/wiki/Europ%C3%A4ische_Patentorganisation, accessed on 26.2.2025.

[78] Wikipedia, https://de.wikipedia.org/wiki/Europ%C3%A4ische_Patentorganisation#Verwaltungsrat, accessed on 26.2.2025.

[79] Federal Patent Court, Cincinnatistraße 64, 81549 Munich.

[80] Federal Patent Court, https://www.bundespatentgericht.de/DE/dasGericht/Aufgaben/aufgaben_node.html, accessed on 27.2.2025.

[81] Federal Patent Court, https://www.bundespatentgericht.de/DE/dasGericht/Organisation/organisation_node.html, accessed on 27.2.2025.

technical inventions, in particular in proceedings on patent granting, nullity proceedings, and utility model cancellation proceedings.[82]

Patent infringement proceedings are not heard before the Federal Patent Court, but rather by certain regional and higher regional courts that have established patent litigation chambers.

1.24 Federal Court of Justice (BGH)

The task of the Federal Court of Justice is, in particular, to ensure uniform case law and to guarantee the further development of the law by the judiciary. The legal opinions of the Federal Court of Justice are generally taken into account by courts in Germany in their decisions.[83]

The Federal Court of Justice currently has 13 civil panels, with the first and tenth panels (I. and X. Civil Panels) responsible for industrial property protection. The I. Civil Panel of the Federal Court of Justice is responsible for legal disputes concerning design law, including Community design law (EU design rights) and trademarks.[84]

The X. Civil Panel of the Federal Court of Justice deals with legal disputes concerning patent and utility model law. In addition, the X. Civil Panel is assigned legal disputes relating to employee invention law and patent nullity cases.[85]

1.25 European Union Intellectual Property Office (EUIPO)

The European Union Intellectual Property Office (EUIPO) is based in Alicante, Spain, and is responsible within the European Union for the administration of EU trademarks and registered Community designs.[86]

---

[82] Federal Patent Court, https://www.bundespatentgericht.de/DE/dasGericht/Organisation/organisation_node.html, accessed on 27.2.2025.

[83] Federal Court of Justice, https://www.bundesgerichtshof.de/DE/DasGericht/Aufgaben/aufgaben_node.html;jsessionid=D068F148432085D210C19514061B1903.internet971, accessed on 27.2.2025.

[84] Federal Court of Justice, https://www.bundesgerichtshof.de/DE/DasGericht/Geschaeftsverteilung/Geschaeftsverteilungsplan2025/Zivilsenate2025/zivilsenate2025.html?nn=10937200#1, accessed on 27.2.2025.

[85] Federal Court of Justice, https://www.bundesgerichtshof.de/DE/DasGericht/Geschaeftsverteilung/Geschaeftsverteilungsplan2025/Zivilsenate2025/zivilsenate2025.html?nn=10937200#10, accessed on 27.2.2025.

[86] EUIPO, https://www.euipo.europa.eu/de/about-us/the-office, accessed on 9.3.2025.

Grant of patents 2

Contents

The patent granting procedure serves to find a claim wording that provides the optimal, that is, the broadest possible scope of protection. The broadest possible scope of protection must be determined in light of the prior art identified in the search. If no claim wording can be found that meets the patentability criteria, the patent application will be rejected.[1]

This chapter presents in particular the German granting procedure. The European patent granting procedure is very similar to the German one. Where significant differences arise, these will be specifically pointed out.

In the German procedure, the patent granting process is initiated by a request for examination.[2] If the request for examination is filed together with or at least shortly after the submission of the application documents, the patent office endeavors to issue the first

[1] Section 48 Patent Act.

[2] Section 44 paragraph 1 Patent Act.

T. H. Meitinger, *Grant of a Patent, Opposition, Appeal, and Invalidity*,
https://doi.org/10.1007/978-3-662-72556-6_2

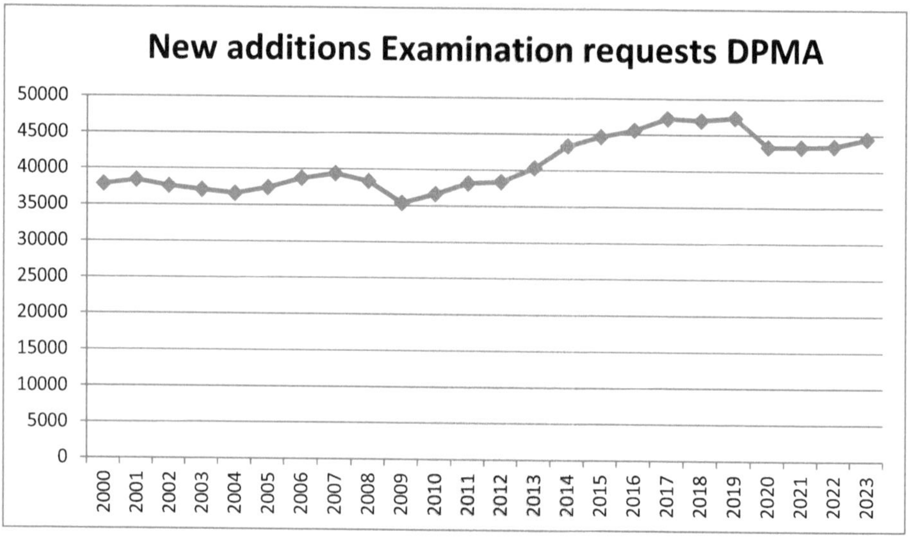

Fig. 2.1 New examination requests received by the GPTO from 2000 to 2023 (DPMA, Annual Reports of the DPMA, https://www.dpma.de/dpma/veroeffentlichungen/jahresberichte/index.html, accessed on 21.12.2024)

official communication within the one-year priority period. This enables the applicant to assess the prospects of success and to make an informed decision as to whether subsequent applications abroad are advisable.

The patent granting procedure takes 3 to 8 years, during which two to three official communications must be answered before allowable claims are present or a rejection is issued by the patent office. A rejection by the patent office can be appealed (see Chap. 5: Applicant's appeal).

The request for examination must be filed within the first seven years after submission of the application documents.[3] The request for examination is only validly filed if an examination fee is paid.[4] Otherwise, the application will be rejected. In the European procedure, no separate request is required. The examination procedure begins automatically.

2.1 Significance of the Patent Granting Procedure

Fig. 2.1 shows the number of examination requests received for the years 2000 to 2023. During this period, the number of examination requests to be processed by the German Patent and Trade Mark Office (GPTO, in German: DPMA, which is the

[3] Section 44 paragraph 2 sentence 1 Patent Act.

[4] Section 44 paragraph 2 sentence 2 Patent Act.

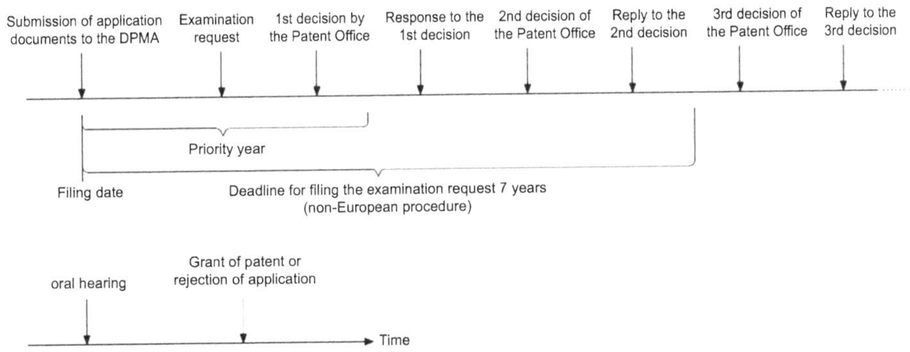

Fig. 2.2 Timeline of a granting procedure

abbreviation for "Deutsches Patent- und Markenamt") increased by approximately 18%.

2.2 Timeline of the Patent Granting Procedure

Fig. 2.2 shows the timeline of the granting procedure before the patent office from the submission of the application documents, which leads to the assignment of a filing date,[5] to the grant of the patent or the rejection of the patent application.[6]

Fig. 2.2 shows that, as a rule, two to three official communications must be answered by the applicant. The examination procedure may be terminated without an oral hearing.[7]

2.3 Formal Examination

After a formal examination by the patent office, a filing date is assigned to the application.[8] The prerequisite for assigning a filing date is that the submitted patent application includes the name of the applicant, a request for the grant of a patent, and a description of the invention.[9] It is not required for a filing date that the application contains patent claims and drawings illustrating the various embodiments of the invention.[10]

---

[5] Section 35 paragraph 1 Patent Act.
[6] Section 42 paragraph 3 sentence 1 Patent Act.
[7] Section 45 paragraph 1 sentence 1 Patent Act.
[8] Section 35 paragraph 1 Patent Act.
[9] Section 35 paragraph 1 in conjunction with Section 34 paragraph 3 nos. 1, 2, and 4 Patent Act.
[10] Section 34 paragraph 3 nos. 3 and 5 Patent Act.

2.4 Inventor Designation

The applicant is required to designate the inventor. For this purpose, a period of 15 months from the filing date is granted for inventor designation or, if priority has been claimed, a correspondingly shortened period is set.[11]

The applicant must indicate how ownership of the invention was transferred to him, if he is not the inventor.[12] For example, a transfer of ownership may occur through the claiming of an employee's invention by the employer on the basis of the Employee Inventions Act.[13]

The inventor must be named in the published application,[14] in the patent specification[15] and in the publication of the grant of the patent.[16]

2.5 Substantive Examination

The substantive examination particularly covers the patentability criteria of novelty, inventive step and industrial applicability. Care must be taken to ensure that the claims are not impermissibly broadened during the grant procedure.

2.6 Response to Office Action Regarding Novelty and Inventive Step

A response to an office action constitutes the reply to an official communication in a patent grant procedure. In the official communication, the patent examiner lists the deficiencies that prevent the grant of a patent. Within a time limit set by the examiner, these deficiencies must be remedied by the applicant. If the time limit is only two months, only formal deficiencies need to be remedied in order to proceed to grant. If the examiner sets a four-month time limit, it can be assumed that the examiner does not yet foresee a prompt grant, as essential points still need to be clarified. For example, in this case, there will not yet be allowable claims that are novel and inventive.

If allowable claims are present, it is usually still necessary to adapt the description to the new claims and to "acknowledge" the cited prior art in the introductory part of the description. For acknowledgment, it is sufficient to briefly mention the document with its official file number and to describe the subject matter of the document in one sentence.

[11] Section 37(1) sentence 1 Patent Act.

[12] Section 37(1) sentence 2 Patent Act.

[13] Section 6(1) Patent Act.

[14] Section 32(2) Patent Act.

[15] Section 32(3) Patent Act.

[16] Section 58(1) Patent Act.

In the office action response, it should first be described which documents are newly submitted and which previous documents are thereby replaced.

For example, the following could be described:

New Documents
New claims 1 to 10 are submitted, replacing the original set of claims.

and/or:

New Documents
New description pages are submitted, replacing the previous description.

If description pages are to be amended, it is advisable to submit a complete new set of description pages rather than only the amended ones. This saves the examiner the time-consuming and error-prone task of sorting the new description pages into the previous set.

If claims are amended, it must be indicated how the new claims are derived from the previous claims and the description. For example, a section in the office action response could read:

Disclosure of the Claims
The subject matter of the new claims is disclosed at the following locations in the original documents:

New claim 1 comprises the features:

- *original claim 1*
- *original claim 5*
- *original claim 6*

New claim 2 comprises the features:

- *original claim 2*
- *original claim 5*
- *The feature "the turbine is arranged directly on the device" can be found on page 2, first paragraph of the originally filed application documents.*

The further claims of the new set correspond to the previous set of claims.
Thus, the subject matter of the new claims does not extend beyond the content of the original application documents.

If the official communication objected to lack of novelty of the previous main claim, for each document of the prior art that challenges novelty, the features of the new main claim that are not disclosed in the prior art document must be identified to establish novelty.

Typically, features are added to a non-allowable main claim from dependent claims or from the description, so that the main claim becomes novel and inventive.

In the novelty examination, it is advisable to briefly explain in advance what the subject matter of the prior art document is. In addition, it is advantageous to use a drawing from the document to show what is not implemented as in the new main claim. For this purpose, a drawing from your own patent application can also be used for comparison. The drawings should be inserted at the relevant points in the office action response to spare the examiner from having to search for the drawings.

For example, the section on novelty could be structured as follows:

> ***Noveltyof claim 1:***
> *D1 (US...) describes a device for milling .... Document D2 (DE...) discloses a device for gluing chipboard elements, wherein .... None of the documents describes a ....*
> *Therefore, the subject-matter of claim 1 is novel, and consequently so are the subject-matters of dependent claims 2 to 10.*

Inventive step is to be addressed in several steps, whereby one can follow the problem-solution approach, which was developed by the European Patent Office.

The problem-solution approach is carried out in three steps. In the first step, the closest prior art is determined. This is, in particular, the document of the prior art that shares the most features with the invention under examination and is therefore the best starting point to arrive at the invention.[17]

In the next step, the objective technical problem to be solved in order to proceed from the closest prior art to the invention is determined. For this purpose, first the features are identified that are part of the invention but are not present in the closest prior art.[18] The objective technical problem is therefore to achieve the effect that results from the distinguishing features.

If no documents of the prior art can be found that, together with the closest prior art, lead to the invention, the invention is inventive. If documents can be found which, in combination with the closest prior art, result in the invention, it must still be examined whether the skilled person would also have considered these documents in order to arrive at the invention. Only if the skilled person would have had motivation and indications to combine the respective documents is the invention actually not inventive. The European Patent Office refers to this assessment as the "Could-Would Test", i.e., whether the skilled person not only could combine the documents (could), but also would combine the documents (would).[19]

Inventive step could be substantiated as follows in a response to an office action:

[17] European Patent Office, https://www.epo.org/de/legal/guidelines-epc/2024/g_vii_5_1.html, accessed on 4.3.2025.

[18] European Patent Office, https://www.epo.org/de/legal/guidelines-epc/2024/g_vii_5_2.html, accessed on 4.3.2025.

[19] European Patent Office, https://www.epo.org/de/legal/guidelines-epc/2024/g_vii_5_3.html, accessed on 4.3.2025.

> ***Inventive step of claim 1***
> *Document D1 represents the closest prior art, since D1 describes a device for ....*

In the first step, the closest prior art is identified and it is explained why this document is the closest. A justification may be that the patent application and the closest prior art pursue the same objective and belong to the same technical field. Alternatively, the common features can be listed, thereby clarifying that the closest prior art and the invention are technically related, and the closest prior art is therefore the most promising document to arrive at the invention.

> *D1 does not disclose that ....*

In the second step, the so-called "distinguishing features" are identified. These are the features of the invention that are not present in the closest prior art.

> *This makes it possible to provide a device that is efficient and fast ....*

In the third step, the effects resulting from the distinguishing features are explained.

> *Objective technical problem is therefore to provide a device that ....*

In the fourth step, the objective technical problem is described, the solution of which leads from the closest prior art to the invention. This problem consists precisely in achieving the effect caused by the distinguishing features.

> *The subject-matter of claim 1 is not obvious over D1 or D2, or over D1 in combination with common general knowledge, or over D2 and common general knowledge, or over a combination of D1 with D2 with or without common general knowledge, since none of the documents pursues such an objective and therefore there are no indications that through the features, that ... this objective can be achieved.*

In the fifth step, it must be clarified that the possible combinations of the prior art with the closest prior art do not lead to the invention, in particular because the skilled person would not combine the documents and/or because, even if the documents were combined, the skilled person would still have to act inventively to arrive at the invention.

> *Therefore, the subject matter of the new claim 1 is based on inventive step, and consequently, so are the subjects of dependent claims 2 to 10.*

In a concluding sentence, it can be stated that the claims dependent on claim 1 also become inventive due to the inventive step of claim 1.

These steps must be carried out for all independent claims. The response to the office action contains requests that clarify the scope of protection sought:

Requests
It is therefore requested that a patent be granted on the basis of the new patent claims 1 to 10.
However, if the new patent application is not considered allowable and a rejection of the application is contemplated, an oral hearing is requested as an auxiliary measure.
The revision and adaptation of the description will be carried out as soon as allowable claims are available.

The response to the office action must be signed. The response to the office action should be accompanied by the set of claims and/or the description pages, if these have been amended, in duplicate as an attachment. On the one hand, the new documents should be submitted in clean copy and, in addition, in a marked-up version in which the changes are indicated in a different color and/or underlined.

Attachments:
New set of claims
(Clean copy and marked-up version)

2.7 Response to the office action due to adaptation of the description

If allowable claims already exist in the patent examination procedure, only the description needs to be adapted to the amended claims. In this context, in particular, the prior art cited by the patent office must be mentioned in the patent application. In addition, the introductory part of the description must be adapted to the new claims and the new claims must be incorporated into the description.

Example:

With reference to the communication dated 3 July 20xx:

New documents
New claims 1 to 6 are submitted, replacing the current set of claims. The set of claims has been provided with reference signs and drafted in two-part form.
In addition, new description pages are submitted, replacing the previous description. In the new description, the prior art has been acknowledged in the usual manner, and the problem to be solved and the support for the claims have been adapted to the new set of claims.

Request
The grant of a patent is therefore requested on the basis of the new patent claims 1 to 6.
(Responses to office actions must be signed)

Attachments:
New set of claims
(Clean copy and marked-up version)

2.8 Example 1: Stilt House

Two real examples of responses to office actions are presented, in which in particular lack of novelty and lack of inventive step are discussed.

The first example deals with a stilt house, i.e., a "house" elevated by steel, wood, or concrete supports.[20]

"European Patent Application No. 17 180 716.7

"Stilt House"
With reference to the communication dated 22 January 2018:

New documents

New claims 1 to 9 are submitted, replacing the original set of claims.

Disclosure of the claims

The subject matter of the new claims is disclosed at the following locations in the original documents:

Claim 1:

- *original claim 1*
- *original claim 2*
- *The feature of the stilt house as a conference room is disclosed, for example, on page 4, second paragraph of the original description documents.*

The further claims of the new set of claims correspond to the previous set of claims.

Thus, the subject matter of the new claims does not extend beyond the content of the original application documents.

Novelty

Claim 1
D1 (US 4719716) describes a mobile shooting range for sports shooters. The shooting range is mounted on a trailer, which can be used as an attachment and is, for example, coupled to a truck. In contrast to new claim 1, no room is disclosed that is built on stilts and can therefore be easily relocated with a truck to serve as a mobile conference room. The disclosure of D2 (US 20130118540 A1) shows an erectable shooting range, and D3 (US 20140083023 A1) shows a protective device. Both documents D2 and D3 do not go beyond the technical teaching of D1.

[20] European Patent Office, https://register.epo.org/application?documentId=E1P6TFG67978DSU&number=EP17180716&lng=de&npl=false, retrieved on 8.3.2025.

Therefore, the subject matter of claim 1 is new, and consequently, so are the subjects of dependent claims 2 to 9."[21]

In a response to an office action, it is advisable to use the drawings from your own patent application and those of the prior art to highlight the differences. In this case, Figures 3 and 4 of US 4,719,716 could have been shown, which depict a fixedly mounted shooting range.

Fig. 2.3 shows Figures 3 and 4 of document US 4719716 of the prior art, in which a shooting range is depicted that is not suitable for mobile use.

Fig. 2.4 shows Fig. 9 of US 2013 0118540 A1 of the prior art, in which the shooting range has feet that are fixedly mounted to the ground.

In contrast, Fig. 2.5 of the applicant's own patent application shows the inventive habitation capsule, which has feet to ensure good stability, but which are not anchored in the ground. Therefore, a truck can be driven under the habitation capsule of the inventive device and load it in order to transport the habitation capsule to another location.

"Inventive Step

Claim 1

Document D1 represents the closest prior art, since D1 discloses a mobile, round, UFO-shaped housing.

D1 does not describe a stilt house, which, due to its arrangement on stilts, can be easily relocated by a truck to serve as a mobile conference room.

This additional feature has the effect that a mobile meeting room can be set up and dismantled at any location in the shortest possible time.

Objective technical problem is therefore to provide a conference room that is easy to transport.

The subject-matter of claim 1 is not obvious in view of D1 or D2, or in view of D1 in combination with common general knowledge, or in view of D2 and common general knowledge, or in view of any other combination of the documents of the prior art with or without common general knowledge, since none of the documents pursues such an objective and therefore there are no indications that this problem can be solved by constructing a conference room on stilts.

Therefore, the subject-matter of new claim 1 is based on an inventive step, and consequently also the subject-matter of dependent claims 2 to 9.

Requests

It is therefore requested that a patent be granted on the basis of the new claims 1 to 9.

However, if the new patent application is not considered allowable and a rejection of the application is contemplated, an oral hearing is requested as an auxiliary measure.

[21] European Patent Office, https://register.epo.org/application?documentId=E1P6TFG67978DSU&number=EP17180716&lng=de&npl=false, retrieved on 8.3.2025.

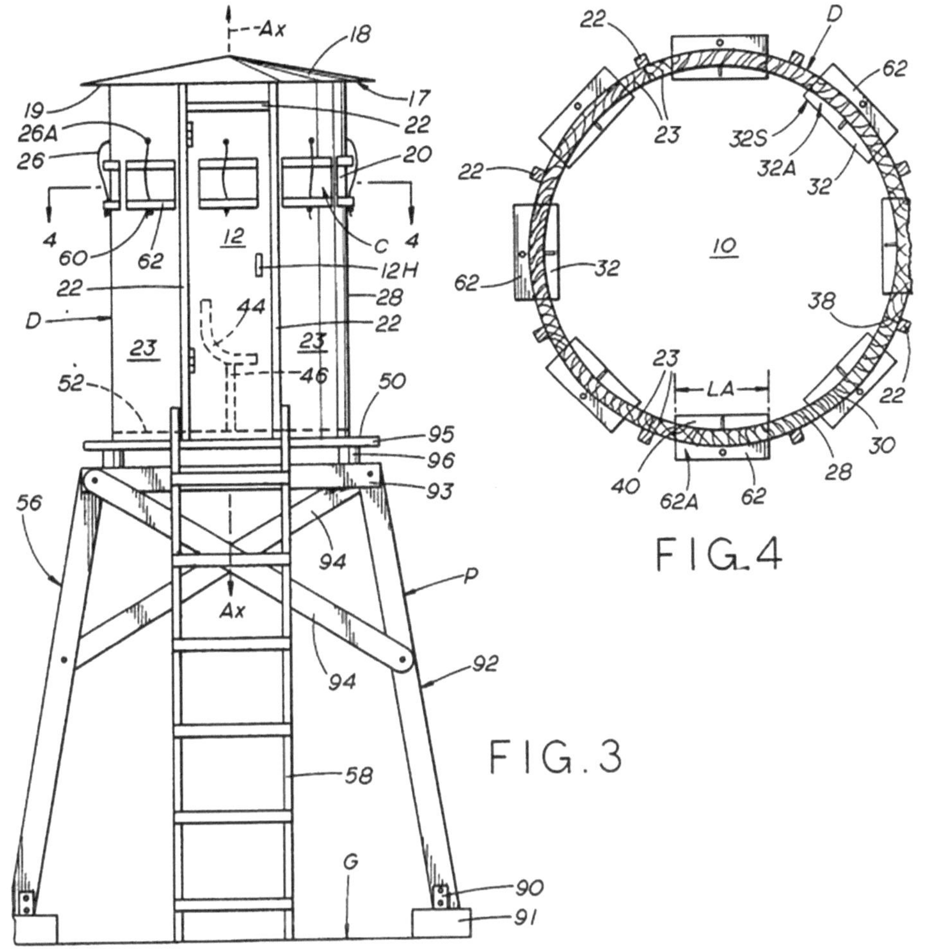

Fig. 2.3 Figs. 3 and 4 of US4719716[22]

The revision and adaptation of the description will be carried out as soon as allowable claims are available. (Responses to office actions must be signed.)

Annexes:
New set of claims
(Clean copy and marked-up copy)"[23]

[22] DPMA, https://depatisnet.dpma.de/DepatisNet/depatisnet?action=pdf&docid=US000004719716A&xxxfull=1, retrieved on 8.3.2025.

[23] European Patent Office, https://register.epo.org/application?documentId=E1P6TFG67978DSU&number=EP17180716&lng=de&npl=false, retrieved on 8.3.2025.

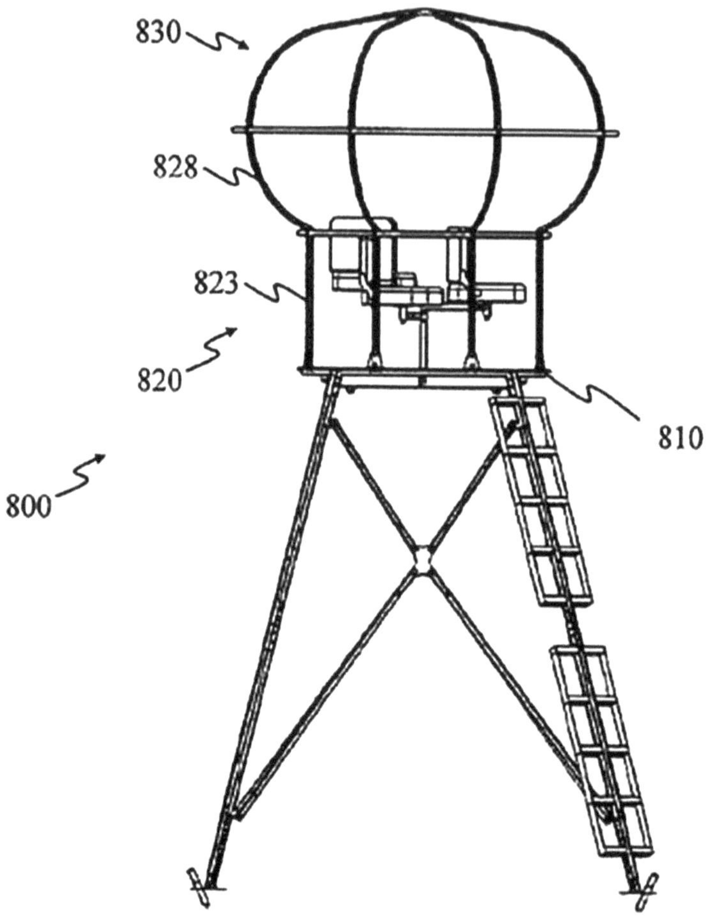

FIG. 9

Fig. 2.4 Fig. 9 of US20130118540A1[24]

An adjustment of the description can already be made at an early stage, which makes it easier for the examiner to consider a prompt grant of the patent. However, if it is foreseeable that a set of claims suitable for grant is not available at the current stage, an adjustment of the description will rather only result in unnecessary workload. If, however, the

[24] DPMA, https://depatisnet.dpma.de/DepatisNet/depatisnet?action=pdf&docid=US020130118540A1&xxxfull=1, retrieved on 8.3.2025.

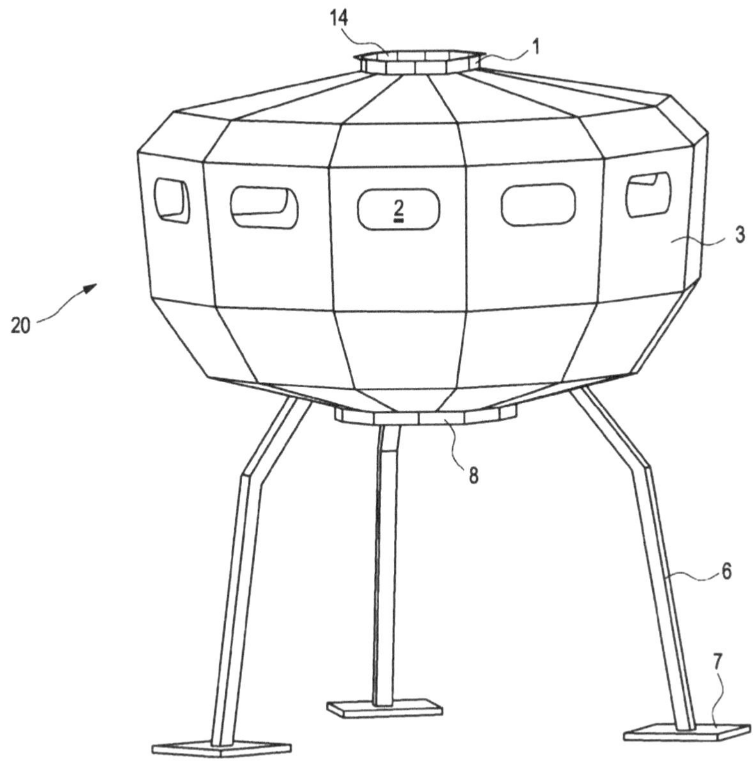

Fig. 2.5 Fig. 8 of EP3269899A1[25]

examiner signals an intention to grant the patent, the applicant should promptly provide the fully corrected application documents to the patent office.

2.9 Example 2: Nail Pictures

The second example of a response to an office action deals with the production of nail pictures.[26] Nail pictures comprise a multitude of nails, which are arranged either far apart or close together, thereby creating a visual impression. Additionally, the nails can be connected with threads, thus also forming lines. Overall, an aesthetic impression is

[25] DPMA, https://depatisnet.dpma.de/DepatisNet/depatisnet?action=pdf&docid=EP000003269899A1&xxxfull=1, retrieved on 8.3.2025.

[26] European Patent Office, https://register.epo.org/application?documentId=M4TNWKWM29ULXOG&number=EP21749531&lng=de&npl=false, accessed on 8.3.2025.

intended, resembling a painting. In the grant procedure of EP 4182106 before the European Patent Office, the following response to an office action was filed.[27]

> *"In response to the communication dated 12 September 2024:*
>
> **New Documents**
>
> *New claims 1 to 4 are filed, which replace the original set of claims.*
>
> **Disclosure of the Claims**
>
> *The subject-matter of the new claims is disclosed at the following locations in the original documents:*
>
> *The new claim 1 comprises the features:*
>
> - *original claim 6*
> - *original claim 1*
> - *original claim 4*
> - *original claim 7*
> - *The feature … can be found in paragraph 4 on page 2 of the originally filed application documents.*
>
> *The new claim 2 comprises the features:*
>
> - *original claim 8*
> - *original claim 1*
> - *original claim 2*
>
> *The further claims of the new set of claims correspond to the previous set of claims.*
>
> *Thus, the subject-matter of the new claims does not extend beyond the content of the original application documents."*[28]

It must always first be explained which new documents are being filed and on which passages in the originally filed application documents the amendments are based. Preferably, the disclosure location is specified for each individual newly introduced feature. In this way, an inadmissible extension can be reliably excluded.

> **"Novelty**
>
> ***Claim 1***
>
> *The D1 (US4964774A) describes a device that can be used for the production of string art, showing nails with a thickening or a head on one side (see Fig. 2.6).*
>
> *Document D2 (US6149436A) also shows nails with a head (see Fig. 2.7).*

---

[27] European Patent Office, https://register.epo.org/application?number=EP21749531&lng=de&tab=doclist, accessed on 13.3.2025.

[28] European Patent Office, https://register.epo.org/application?documentId=M4TNWKWM29ULXOG&number=EP21749531&lng=de&npl=false, accessed on 8.3.2025.

Fig. 2.6 Figs. 1 and 3 of US4964774A[29]

FIG. 1

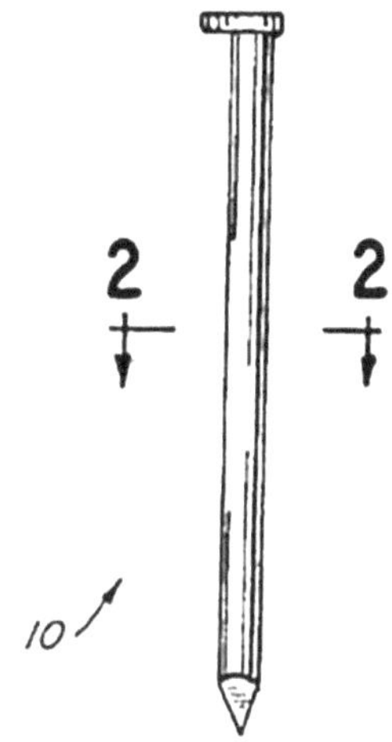

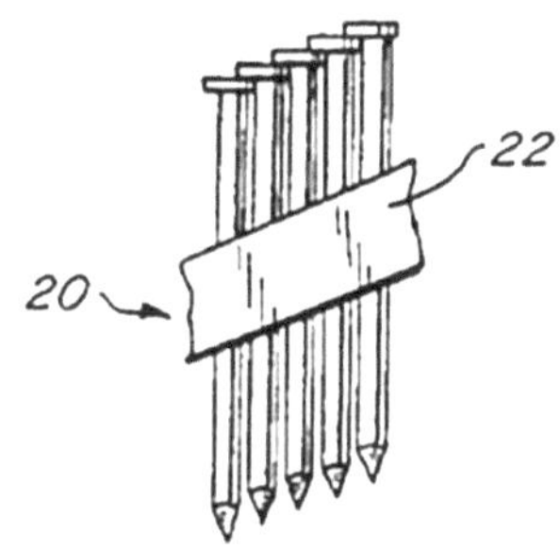

FIG. 3

Document D3 (US1440579A) describes holding devices that have a thickened portion at one end for receiving a thread (see Fig. 2.8). The nails of D3 are intended for the manufacture or repair of shoes.

Fig. 2.9 shows chains of nail blanks.

D3 and D4 show nail chains, in which these have a structure as an endless wire.

In the subject matter of the current main claim, only an endless wire is mentioned. Figure 1 of the present patent application shows the endless wire, in which no structure can be identified. The skilled person will therefore, in the subject matter of the current claim,

---

[29] DPMA, https://depatisnet.dpma.de/DepatisNet/depatisnet?action=pdf&docid=US000004964774A&xxxfull=1, accessed on 13.3.2025.

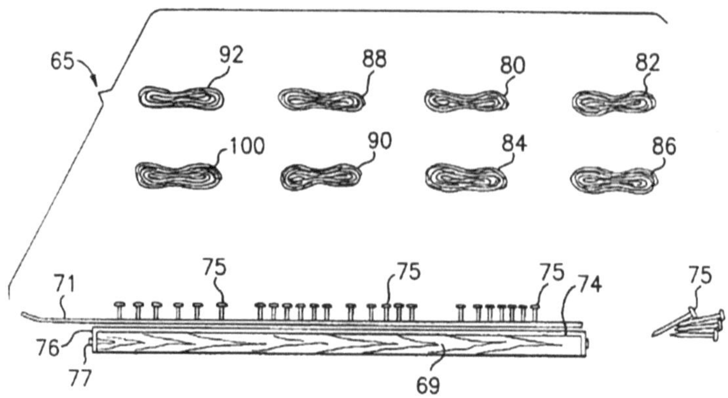

Fig. 2.7 Fig. 1 of US6149436A[30]

> *assume a simple, standard endless wire that is identical and without contour along its entire length (Fig. 2.10).*
>
> *Therefore, in none of the documents of the prior art is an endless wire described that is intended for the production of string art and shows headless nails with a tip for easier driving into a board. Therefore, the subject matter of claim 1 is new, and consequently also the subject matters of claims 2 to 4."*[31]

By including drawings in the response to the office action, it can make it easier for the examiner to understand the arguments presented.

> **"Inventive Step**
>
> ***Claim 1***
>
> *Document D2 represents the closest prior art, as D2 describes a string art representation. Documents D1, D2, and D3 do not show a string art representation and therefore would not be used by the skilled person to improve the manufacture or the string art representation itself. In addition, documents D3 and D4 are also affected by their age, as D3 was filed at the patent office in 1920 and D4 even as early as 1885. The skilled person will always try to apply the current technology, as this is typically the most advanced. D2 is therefore the closest prior art for both claim 1 and claim 2, since both claims fall within the technical field of string art representation.*
>
> *D2 does not disclose that the nails for the string art representation can be provided using a simple, contourless endless wire, where the end of the endless wire is driven into the board and then the endless wire is severed in such a way that the new end of the endless wire forms a point, so that it can easily be driven into the board at another location.*

---

[30] DPMA, https://depatisnet.dpma.de/DepatisNet/depatisnet?action=pdf&docid=US000006149436A&xxxfull=1, accessed on 13.3.2025.

[31] European Patent Office, https://register.epo.org/application?documentId=M4TNWKWM29ULXOG&number=EP21749531&lng=de&npl=false, accessed on 8.3.2025.

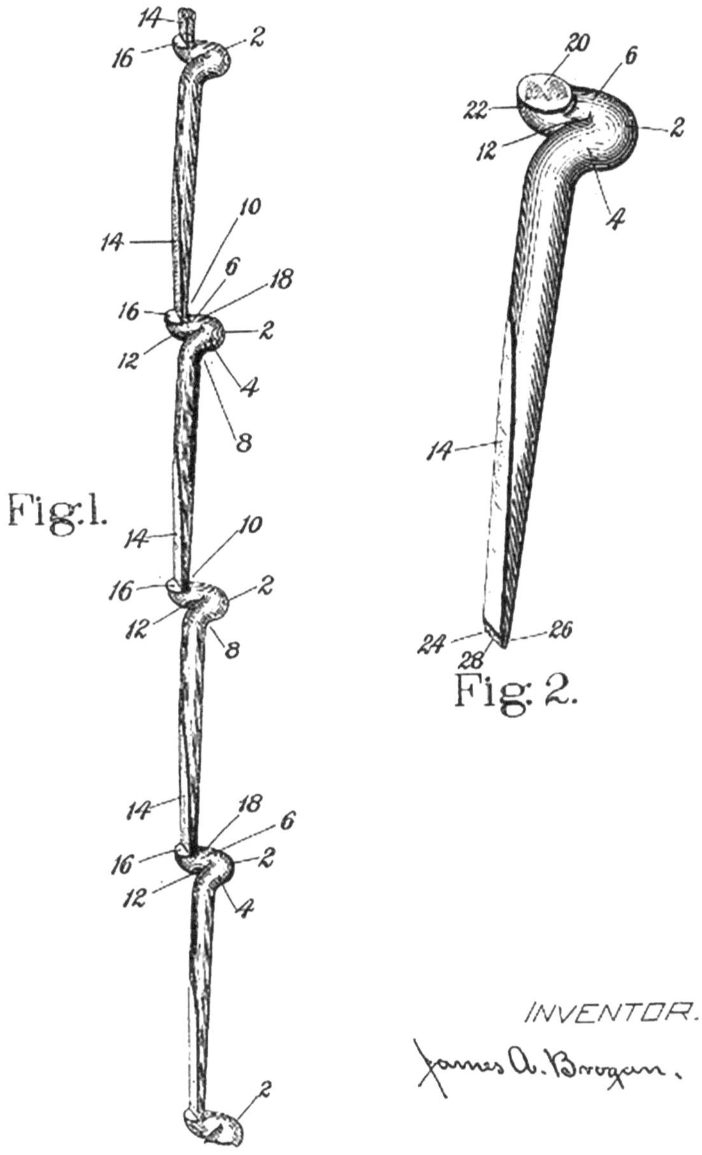

Fig. 2.8 Figs. 1 and 2 of US1440579A[32]

[32] DPMA, https://depatisnet.dpma.de/DepatisNet/depatisnet?action=pdf&docid=US000001440579A&xxxfull=1, accessed on 13.3.2025.

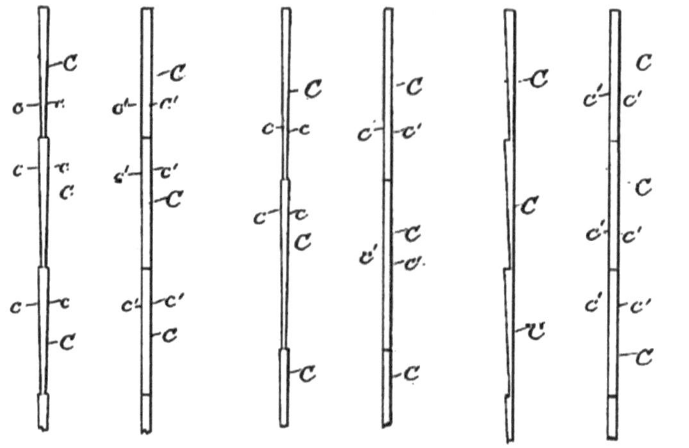

Fig. 2.9 Figs. 1 to 6 of US315070[33]

Therefore, manufacturing string art in a very short time with high quality is not possible in the prior art (see page 2, lines 22 to 24 and page 5, lines 10 to 19 of the filed application documents).

The objective technical problem is therefore to provide a device with which string art can be carried out very quickly and cost-effectively.

Document D1 deals with fastening elements made of metal. Document D3 describes a chain of nails used in the manufacture of shoes. Document D4 describes nail blanks. None of documents D1, D3, or D4 provide any indication that the technical teachings of these documents could be used to optimize the manufacture of a string art representation. In particular, D1, D3, and D4 describe manufacturing methods for nails, but do not disclose or suggest driving the unfinished nail into a board before completing the final nail manufacturing process step. Nor does the closest prior art provide any indication that, for example, an endless wire or the teachings of documents D1, D3, or D4 could be used to achieve optimization. The skilled person would therefore not combine D2 with D1, D3, or D4.

Therefore, the subject matter of new claim 1 is based on an inventive step. The arguments for claims 2 to 4 are analogous."[34]

The argumentation of inventive step must be carried out for each independent claim.

[33] DPMA, https://depatisnet.dpma.de/DepatisNet/depatisnet?action=pdf&docid=US000000315070A&xxxfull=1, accessed on 13.3.2025.

[34] European Patent Office, https://register.epo.org/application?documentId=M4TNWKWM29ULXOG&number=EP21749531&lng=de&npl=false, retrieved on 8.3.2025.

Fig. 2.10 Fig. 1 of EP4182106[35]

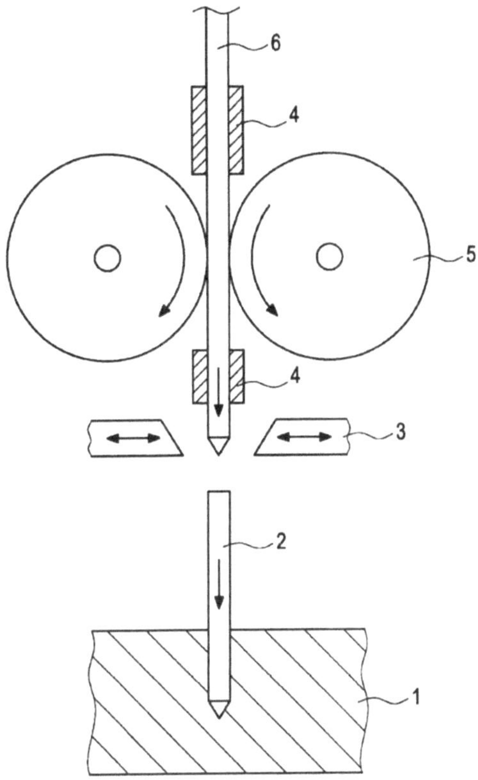

Fig. 1

"Requests

It is therefore requested that a patent be granted on the basis of the new patent claims 1 to 4.

However, if the new patent application is not considered allowable and a rejection of the application is contemplated, oral proceedings are requested as an auxiliary measure.

The revision and adaptation of the description will be carried out as soon as allowable claims are available."[36]

[35] European Patent Office, https://register.epo.org/application?documentId=E6EYDYRF7212DSU&number=EP21749531&lng=de&npl=false, accessed on 13.3.2025.

[36] European Patent Office, https://register.epo.org/application?documentId=M4TNWKWM29ULXOG&number=EP21749531&lng=de&npl=false, retrieved on 8.3.2025.

It is important that a request for oral proceedings is filed. In this case, the examiner must give the applicant an opportunity to justify the patent application in oral proceedings before rejecting the patent application.

German Opposition Procedure 3

Contents

An opposition is a procedure that allows any third party to have the patentability reviewed in a contentious proceeding at low cost. The third party does not have to demonstrate a legitimate interest. The opposition is subsequent to the grant of the patent, so the patent owner is not prevented from enforcing their patent by an opposition.

A patent division of the German Patent Office is responsible for handling an opposition.[1]

[1] Section 61(1) sentence 1 of the Patent Act.

T. H. Meitinger, *Grant of a Patent, Opposition, Appeal, and Invalidity*,
https://doi.org/10.1007/978-3-662-72556-6_3

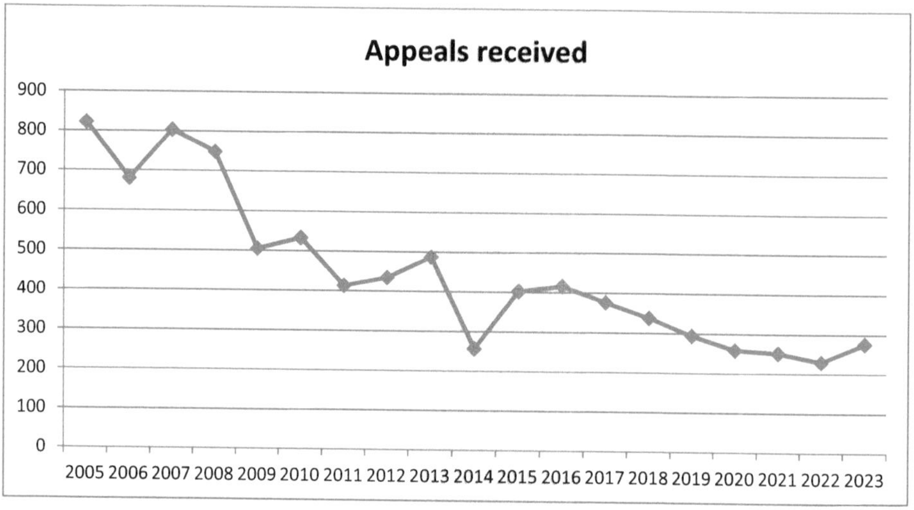

Fig. 3.1 Number of oppositions from 2010 to 2023[2]

3.1 Significance of the Opposition Procedure

Figure 3.1 shows that the number of oppositions filed with the DPMA decreased from 822 in 2005 to 276 new filings in 2023. The number of oppositions has therefore dropped to a third. Currently, significantly fewer oppositions are being filed with the German Patent Office compared to previous years. This development is not due to a substantial decrease in the number of applications, as in the same period from 2005 to 2023, German patent applications only declined from 60,222 to 58,656, i.e., by about 2.6% (see Fig. 3.2).

3.2 Timeline of an Opposition Procedure

Figure 3.3 illustrates the chronological sequence of an opposition procedure, where an opposition can be filed within a period of nine months after publication of the grant of the patent. It is common practice for the patent proprietor to file auxiliary requests with amended sets of claims prior to oral proceedings.

[2] DPMA, Annual Reports of the DPMA, https://www.dpma.de/dpma/veroeffentlichungen/jahresberichte/index.html, accessed on 21.12.2024; in the period from 2001 to 2006, oppositions were not heard before the DPMA, but before the Federal Patent Court. These opposition proceedings are included here.

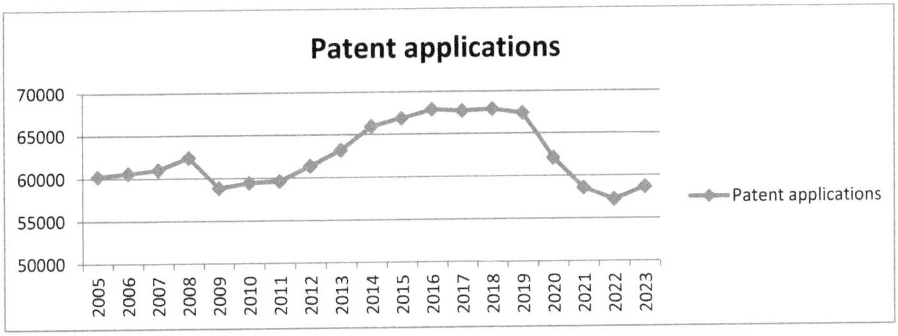

Fig. 3.2 Number of patent applications from 2005 to 2023[3]

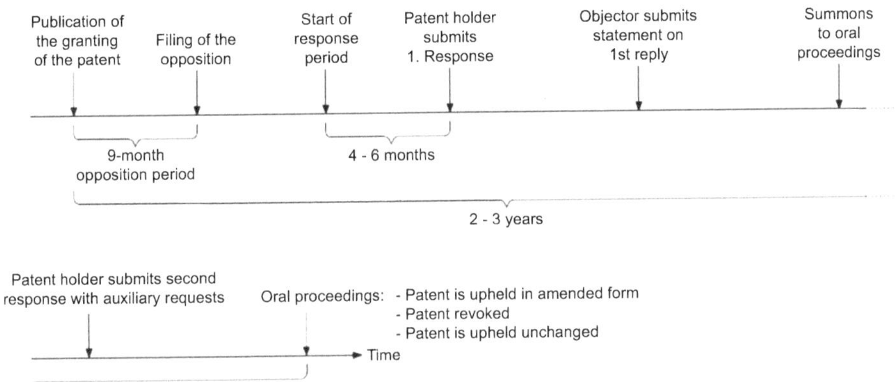

Fig. 3.3 Timeline of an opposition procedure

3.3 Court-like Administrative Procedure

The opposition procedure before the DPMA is neither a purely administrative procedure nor a judicial procedure. The opposition procedure is similar to the nullity proceedings before the Federal Patent Court, as in both cases a granted patent can be challenged on the same grounds for revocation. The opposition procedure is a hybrid and can be regarded as a court-like administrative procedure.[4]

[3] DPMA, Annual Reports of the DPMA, https://www.dpma.de/dpma/veroeffentlichungen/jahresberichte/index.html, accessed on 21.12.2024.

[4] Federal Court of Justice GRUR 07, 859 Information Transmission Procedure I, X ZB 9/06, judgment of April 17, 2007.

3.4 Legal Status of the Parties Involved

Opponents and patent proprietors are not parties, as in judicial proceedings, but "participants".[5] The participants are not "masters of the proceedings" as would be the case in a court procedure. The withdrawal of an opposition therefore does not necessarily terminate the opposition proceedings, but only the participation of the opponent.[6] Furthermore, the Patent Office may also examine grounds for revocation that were not raised by the opponent.

This does not apply to the ground for revocation of vindication. If the only ground for revocation is vindication, the withdrawal of the opposition terminates the proceedings.

3.5 Principle of Official Investigation

In opposition proceedings, as in the patent granting procedure, the principle of official investigation applies.[7] The opposition division is therefore not bound by the grounds for revocation asserted by the opponent. The reasoning is that the opposition division is not only obligated to the participants in the opposition proceedings, but especially to the public, which has an interest in the revocation of invalid patents. For this reason, the opposition proceedings may also be continued by the opposition division even if the opposition is withdrawn. This applies in particular if the proceedings are ready for decision.

3.6 Course of Proceedings

An opposition procedure generally consists of a written preliminary procedure and a hearing, which corresponds to an oral hearing. In the written preliminary procedure, the participants can exchange their arguments. After the written procedure ends, a hearing may be scheduled. However, a hearing only takes place if at least one of the participants has requested it or if the patent division considers it expedient.[8]

With the summons to the hearing, the patent division informs the participants of the points that are to be clarified in particular during the hearing.[9] It is strongly recommended to address the points raised by the patent division during the hearing.

---

[5] Federal Court of Justice GRUR 95, 333 Aluminium Trihydroxide; Federal Court of Justice GRUR 96, 42 Light Spot.

[6] Section 61(1) sentence 2 of the Patent Act or Rule 84(2) sentence 2 EPC.

[7] Federal Court of Justice GRUR 2008, 87—Change of Patent Proprietor in Opposition Proceedings.

[8] Section 59(3) sentence 1 of the Patent Act.

[9] Section 59(3) sentence 2 of the Patent Act.

3.7 Admissibility of the Opposition

The opposition can be filed with the DPMA within a period of 9 months after publication of the grant of the patent.[10] Within the same period, the opposition must be substantiated, meaning that the grounds for revocation on which the patent is to be revoked must be asserted. In addition, the facts and evidence supporting the grounds for revocation must be indicated.[11]

3.8 Merits of the Opposition

The opposition must be based on grounds for revocation that are founded on facts submitted within the opposition period.[12] Only the following grounds for revocation are admissible in opposition proceedings.[13] In opposition proceedings, lack of novelty, lack of inventive step, lack of industrial applicability, insufficiency of disclosure, inadmissible extension, and vindication/unlawful extraction may be asserted.

Novelty

The subject matter of the independent claims must be novel for a patent to be legally valid.[14] Novelty exists if the claim does not form part of the state of the art. The state of the art comprises all disclosures made available to the public before the filing date. If the patent application claims a priority, the filing date is replaced by the priority date.[15] A disclosure of the state of the art may also occur through public use or presentation.

Inventive Step

An inventive step is present if the invention is not obvious to a person skilled in the art.[16] In Section 4 sentence 1 of the Patent Act, the skilled person is explicitly named as the instance for assessing inventive step.[17] The notional skilled person in patent law possesses average skill and knowledge in the field.

A skilled person will generally apply a technical teaching using their expertise and skills, so that the technical teaching is used by each skilled person in different ways.

---

[10] Section 59(1) sentence 1 of the Patent Act.

[11] Section 59(1) sentences 4 and 5 of the Patent Act.

[12] Section 59(1) sentences 4 and 5 of the Patent Act.

[13] Section 59(1) sentence 3 of the Patent Act.

[14] Section 21(1) no. 1 in conjunction with Section 3(1) sentence 1 of the Patent Act.

[15] Section 3(1) sentence 2 of the Patent Act.

[16] Section 4 sentence 1 of the Patent Act.

[17] Section 4 sentence 1 of the Patent Act: "An invention shall be considered as involving an inventive step if, having regard to the state of the art, it is not obvious to a person skilled in the art."

These variations of the state of the art must not be monopolized, so as not to hinder the everyday work of professionals and the gradual "normal" development of technology. Therefore, only those technical teachings that represent a significant advance over the state of the art can be patented. Only technical teachings that are not obvious to a skilled person can be granted a patent.[18]

Could-Would Test

The skilled person is required for the assessment of inventive step, with the Could-Would Test being relevant.

The skilled person does nothing without a reason.[19] Accordingly, it is not sufficient that a skilled person could have arrived at the invention under review by any arbitrary combination of documents from the state of the art. The skilled person must also have had a reason or motivation to combine the relevant documents.[20]

The Could-Would Test is intended in particular to prevent an ex post assessment, in which every invention appears obvious in hindsight.[21] The notional skilled person in patent law is intended to provide a perspective prior to the creation of the invention, when the technical teaching of the invention was still unknown.[22]

Industrial Applicability

Lack of industrial applicability is a ground for opposition.[23] However, it is unlikely that this ground for revocation will succeed before the Patent Office, as it is hardly conceivable that a technical device would not be usable in some way for industrial purposes.

---

[18] Ann, Section 18. Inventive Achievement in Ann, Patent Law, 8th edition 2022, para. 2.

[19] Federal Court of Justice GRUR 2010, 407 single-piece eyelet; Federal Patent Court 2.2.2016 4 Ni 29/14 (EP); Federal Patent Court 28.11.2017 6 Ni 32/16 (EP); Fitzner/Lutz/Bodewig para. 38.

[20] Benkard PatG/Asendorf/Schmidt/Tochtermann, 12th ed. 2023, PatG Section 4 para. 86.

[21] EPO T 47/91; Singer/Stauder/Luginbühl Art 56 EPC para. 56; on its inadmissibility Federal Court of Justice GRUR 2001, 232 hole punch.

[22] EPO T 2/83 OJ EPO 1984, 265=GRUR Int 1984, 527 Simethicone tablet; EPO T 90/84; EPO T 124/84 EPOR 1986, 297; EPO T 223/84 EPOR 1986, 67; EPO T 256/84; EPO T 265/84 EPOR 1987, 193; EPO T 7/86 OJ EPO 1988, 381=GRUR Int 1989, 226 Xanthine; EPO T 392/86; EPO T 219/87; EPO T 274/87 EPOR 1989, 207; EPO T 564/89; EPO T 274/87 EPOR 1989, 207; EPO T 61/90; EPO T 513/90 OJ EPO 1994, 154=GRUR Int 1994, 618 f. foamed bodies; EPO T 597/92 OJ EPO 1996, 135=GRUR Int 1996, 814 rearrangement reaction; EPO T 167/93 OJ EPO 1997, 229=GRUR Int 1997, 742 bleaching agent; EPO T 203/93; EPO T 406/98; Federal Patent Court 16.7.1997 20 W (pat) 64/95; Schulte para. 58 ff.; Szabo Mitt 1994, 225, 233 f.; Knesch VPP-Rdbr 1994, 70, 72.

[23] Section 59(1) sentence 3 in conjunction with Section 21(1) no. 1 in conjunction with Section 5 of the Patent Act.

Sufficiency of Disclosure

In the question of sufficiency of disclosure, the notional skilled person in patent law serves as the standard of assessment. The technical teaching of a patent is therefore not addressed to a technical layperson, but to a skilled person in the technical field of the invention. Only if, despite the knowledge and skills of an average skilled person, a technical teaching of a patent cannot be reproduced, can the ground for revocation of insufficiency of disclosure be successful.

Inadmissible Extension

The ground for revocation of inadmissible extension exists if the disclosure of the patent goes beyond what was originally disclosed at the Patent Office.[24]

3.9 Interpretation of Claims

The independent claims of a patent determine its scope of protection.[25] The independent claims therefore constitute the central subject of negotiation in any contentious proceedings.[26]

A claim requires interpretation in order to correctly determine its meaning. A claim can be interpreted literally, systematically, and functionally. The interpretation of a claim must always be carried out in the context of the entire disclosure of the patent.[27]

It must be taken into account during interpretation that the interpretation is made from the perspective of the skilled person, so that only technically meaningful embodiments are included within the scope of protection of the claim. Technically nonsensical embodiments, which could always be included in the scope of protection of a claim, are disregarded.[28]

The interpretation of a claim can be a demanding task. The final authority in this regard is the competent court.[29] It is entirely possible that an initial interpretation by a lower court may be overturned and corrected by the appellate or supreme court. Some

[24] Section 59(1) sentence 3 in conjunction with Section 21(1) no. 4 of the Patent Act.

[25] Section 14 Patent Act.

[26] Nägerl, Section 4. Delimitation from the State of the Art—Novelty and Inventive Step, Haedicke/Timmann, Handbook of Patent Law, 2nd edition 2020, para. 56.

[27] Section 14, sentence 2 Patent Act.

[28] Timmann, Section 3. Interpretation and Scope of Protection of Patents in Haedicke/Timmann, Handbook of Patent Law, 2nd edition 2020, paras. 51, 52.

[29] Ann, Section 32. The Scope of Protection of the Patent and Utility Model, Ann, Patent Law, 8th edition 2022, para. 51.

decisions of the Federal Court of Justice are based precisely on a different interpretation of the claims compared to the judgments of the lower courts.[30]

The interpretation of the claims must satisfy the requirements of legal certainty and appropriateness.[31] Appropriateness is present if a revolutionary patent is granted a broader scope of protection than a "trivial patent." However, the interpretation must not go beyond the subject matter described in the claim in order to maintain legal certainty.[32]

Literal Interpretation

A literal interpretation is also referred to as an interpretation according to the wording. In this approach, the meaning of each individual word of the claim is determined. The sum of the meanings of the words yields the meaning of the entire claim.

Systematic Interpretation

In the systematic interpretation of a claim, particular consideration is given to the description of the patent. The description cannot determine the interpretation of a claim or even create its own scope of protection. However, the description must always be taken into account when interpreting a claim. The interpretation of a claim includes at least a literal and a systematic interpretation.[33]

Functional Interpretation

In functional interpretation, the focus is on the problem that the invention is intended to solve, and the claim is interpreted in light of the embodiments suitable for solving this problem.

3.10 Right to Be Heard

Observance of the right to be heard means that the decision of the opposition division may only be based on circumstances regarding which the parties have had an opportunity to comment.[34] The legal principle of the right to be heard is enshrined in the Basic Law.[35] The right to be heard means that surprise decisions based on previously unmentioned circumstances are excluded.

[30] Gröning: The Interpretation of Patent Claims by the Federal Court of Justice—Time and Again a Turning Point in Patent Infringement Proceedings, GRUR 2021, 1477.

[31] Osterrieth, Part 6. Patent Infringement, Osterrieth, Patent Law, 6th edition 2021, para. 856.

[32] BGHZ 100, 249 (254)—Broadcasting Transmission System.

[33] Section 14, sentence 2 Patent Act.

[34] Federal Court of Justice GRUR 1966, 583—Sampling Method; Federal Court of Justice GRUR 1978, 99, 100—Direct Current Transfer Supply.

[35] Article 103(1) Basic Law.

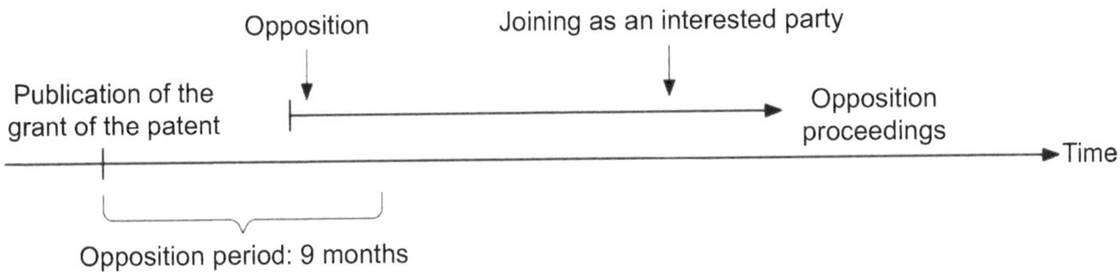

Fig. 3.4 Intervention in opposition proceedings

The parties to the opposition proceedings must therefore be heard on all factual and legal considerations that are relevant to the decision. However, this does not mean that the parties must be informed in advance of how the opposition division will decide.[36]

The right to be heard is, for example, violated if the opponent does not receive new requests from the patent proprietor or if the opponent was not aware of the circumstances that led to the inadmissibility of the opposition.[37]

3.11 Intervention in Opposition Proceedings

Intervention in opposition proceedings (third-party intervention) is available to anyone against whom infringement proceedings based on the contested patent are pending. A three-month intervention period from the filing of the infringement action must be observed.[38] The same applies to a third party who has been requested by the patent proprietor to cease an alleged patent infringement and has subsequently filed a declaratory action for non-infringement of the patent.[39] The opposition of an intervener must be substantiated within the three-month intervention period (Fig. 3.4).[40]

3.12 Withdrawal of the Opposition

The withdrawal of the opposition by an opponent does not mean the termination of the opposition proceedings, even if the opponent is the only one.[41] The withdrawal only ends the status of the opponent as a party to the proceedings. This is appropriate, as the patent office should strive to avoid burdening competition with patents that are not legally valid.

[36] Federal Court of Justice GRUR 1966, 583—Sampling Method.

[37] Federal Patent Court BIPMZ 1986, 181.

[38] Section 59(2) sentence 1 Patent Act.

[39] Section 59(2) sentence 2 Patent Act.

[40] Section 59(2) sentence 3 Patent Act.

[41] Section 61(1) sentence 3 Patent Act.

However, this does not apply if the only ground for revocation raised is vindication/unlawful removal. In this case, withdrawal of the opposition leads to the termination of the opposition proceedings.[42] In the case of vindication, the patent division relies on the cooperation of the opponent to prove the unlawful removal. Furthermore, there is little change for competition if the relevant invention is claimed by another patent proprietor. It is therefore appropriate to terminate the proceedings if the only person with a legal interest does not wish to continue them.

3.13 Decision of the Patent Division

The Patent Division decides on the opposition by means of a decision.[43] The contested patent may be partially revoked, fully revoked, or maintained without amendment.[44]

3.14 Costs of the Opposition Proceedings

As a rule, each party bears its own costs. The opposition proceedings are intended to provide a fast procedure with low cost risk in order to eliminate patents that are not legally valid.

However, the Patent Division may, at its equitable discretion, decide that one of the parties shall bear the costs of the proceedings or of the oral hearing.[45] This may be particularly appropriate if the opposition is withdrawn by the opponent or if the patent proprietor surrenders the patent.[46] The costs of a party may only be imposed on the opposing party to the extent that they were necessary for the protection of the rights and claims of the party.[47] The Patent Division may also determine that the opposition fee is to be refunded if equity so requires.[48]

[42] Section 61(1) sentence 4 Patent Act.
[43] Section 61 paragraph 1 sentence 1 of the Patent Act.
[44] Section 61 paragraph 1 sentence 2 of the Patent Act.
[45] Section 62 paragraph 1 sentence 1 of the Patent Act.
[46] Section 62 paragraph 1 sentence 2 of the Patent Act.
[47] Section 62 paragraph 2 sentence 1 of the Patent Act.
[48] Section 62 paragraph 1 sentence 3 of the Patent Act.

4 Appeal Proceedings Before the Federal Patent Court (BPatG)

Contents

An appeal procedure allows for a review of a decision of the DPMA before the Federal Patent Court (BPatG).[1] In particular, the rejection of a patent application (applicant's appeal) or the decision on an opposition (opposition appeal) can be challenged by means of an appeal.

The appeal procedure takes place before the Federal Patent Court. Nevertheless, the appeal must be filed with the German Patent Office. The reason for this is that the appeal is first forwarded to the examining division or the patent department that issued the contested decision. This gives the examining division or the patent department the opportunity to amend its decision if it considers the appeal to be justified.[2] If the appeal is not "remedied," the appeal is forwarded to the Federal Patent Court without any comment from the Patent Office.[3]

[1] Section 73(1) Patent Act.

[2] Section 73(3) sentence 1 Patent Act.

[3] Section 73(3) sentence 3 Patent Act.

T. H. Meitinger, *Grant of a Patent, Opposition, Appeal, and Invalidity*,
https://doi.org/10.1007/978-3-662-72556-6_4

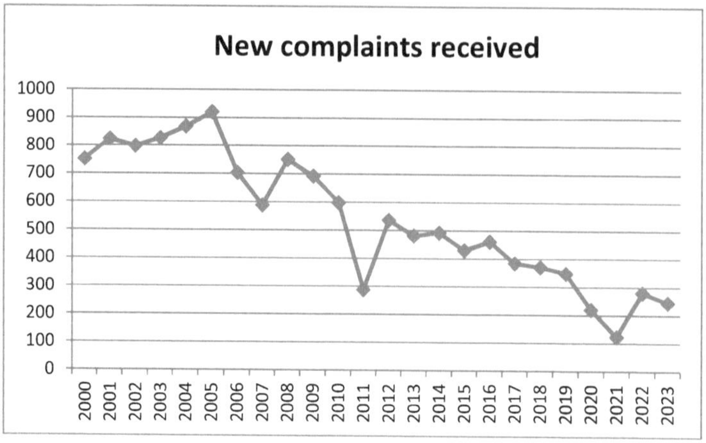

Fig. 4.1 New filings of appeals from 2000 to 2023[4]

4.1 Significance of the Appeal Procedure

Figure 4.1 shows the annual new filings of appeal proceedings at the Federal Patent Court. Opposition appeals are not included in this representation. The number of annual appeals has decreased by approximately 67% from 2000 to 2023.

4.2 Timeline of a German Appeal Procedure

Figure 4.2 shows the timeline of an appeal procedure, which can be requested within one month after service of a decision by the Patent Office.[5]

4.3 Grievance

An appeal is only admissible if the appellant is aggrieved.[6] A grievance exists if the appellant is adversely affected by the contested decision because the Patent Office did not decide in accordance with the appellant's request. Therefore, there must be an objec-

[4] Federal Patent Court, Annual Reports from 2014 to 2023, https://www.bundespatentgericht.de/DE/Presse/Publikationen/Bilderstrecke_Jahresbericht.html?nn=195898&cms_gcp_225314=0#gcp_anchor_225314, accessed on 21.12.2024. or Library of the Federal Patent Court, Annual Reports from 2000 to 2009, XE 180 ac 1/2000 to XE 180 ac 1/2009.

[5] Section 73(2) sentence 1 Patent Act.

[6] BGH GRUR 1972, 535, 536—Lifting of secrecy.

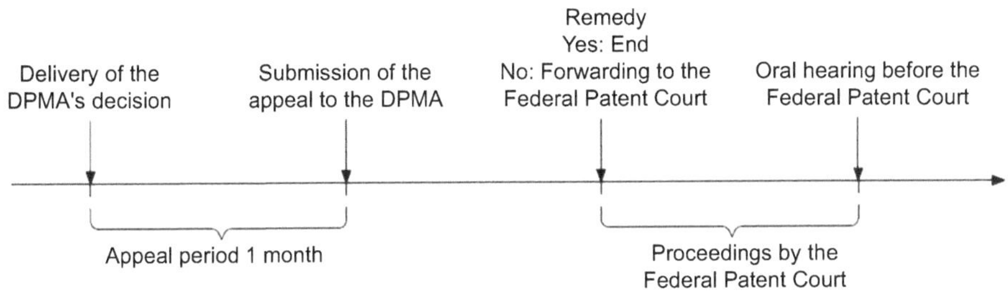

Fig. 4.2 Timeline of a German appeal procedure

tive interest of the appellant in amending the contested decision. A grievance already exists if the decision was made not according to the main request but according to an auxiliary request. An applicant is also aggrieved if the application documents were amended without their consent.[7]

If, as a result of a decision by an opposition division, a patent is only partially maintained, the opponent is aggrieved if they pursued the complete revocation of the patent as their main request. The patent proprietor is aggrieved if they requested the maintenance of the entire patent. Furthermore, the opponent is aggrieved if the patent is maintained, and the patent proprietor is aggrieved if the patent is revoked.

4.4 Admissibility

An appeal is only possible against decisions that finally determine a matter. "Interlocutory decisions" cannot be challenged.[8]

4.5 Permissibility

A permissible appeal exists if an appealable decision is challenged in the correct form and within the appeal period of one month after service of the decision.[9]

[7] BGH GRUR 1982, 291, 292—Polyesterimides; BPatG GRUR 1983, 366, 367.

[8] BPatGE 2002, 56; BPatGE 2010, 35, 39: BPatGE 2013, 163, 164.

[9] Section 73(2) sentence 1 Patent Act.

The appeal must be filed in writing. The appellant must be identifiable from the appeal.[10] If the appellant cannot be clearly identified, the appeal is inadmissible.[11] A correction after expiry of the appeal period is irrelevant.

4.6 Merits

The appeal must be substantiated, meaning that all facts and evidence supporting the appeal must be presented.[12]

4.7 Waiver of Appeal

A waiver of appeal may be declared. If an appeal is nevertheless filed in due time and form, the appeal is deemed inadmissible.[13]

[10] BGH 12.10.1976, X ZB 18/74, GRUR 1977, 508—Intercepting device; BGH 7.11.1989, X ZB 24/88, GRUR 1990, 108, 109—Measuring head.

[11] BGH GRUR 1977, 508.

[12] Section 73(3) sentence 1 Patent Act.

[13] BPatGE 2015, 153.

German Applicant's Appeal 5

Contents

A rejection decision by an examination division can be appealed.[1] The appeal must be filed with the DPMA within one month of service of the contested decision.[2] The appeal has a suspensive effect.[3] This does not apply if the invention is a state secret and the Patent Office has prohibited publication of the invention.[4]

[1] Section 73(1) Patent Act.

[2] Section 73(2) sentence 1 Patent Act.

[3] Section 75(1) Patent Act.

[4] Section 75(2) in conjunction with Section 50(1) sentence 1 Patent Act.

T. H. Meitinger, *Grant of a Patent, Opposition, Appeal, and Invalidity*,
https://doi.org/10.1007/978-3-662-72556-6_5

5.1 Remedy

The appeal is forwarded to the examination division that issued the contested decision. If the examination division considers the appeal to be justified, it corrects its decision and thus grants the appeal.[5] In addition to granting the appeal, the examination division may also order the refund of the appeal fee.[6]

If the appeal is not granted, the appeal is forwarded to the Federal Patent Court. No comment or statement from the examination division, the patent department, or the Patent Office is attached to the appeal.[7]

5.2 Participation of the Patent Office

Since appeals are directed against the Patent Office, the President of the Patent Office is given the opportunity to submit written statements, attend the oral proceedings, and speak during them. However, the participation of the President of the Patent Office is only permissible if it is required to safeguard the public interest.[8] In addition, the President of the Patent Office may submit written statements to the Federal Patent Court, which must be communicated to the parties.[9]

If the appeal concerns a matter of fundamental importance, the Federal Patent Court may invite the President of the Patent Office to join the appeal proceedings.[10] By joining the appeal proceedings, the President of the Patent Office becomes a party to the appeal proceedings.[11]

5.3 Oral Proceedings

As a rule, the appeal proceedings include an oral hearing before the Federal Patent Court. The prerequisite for this is that the appellant files a request for an oral hearing.[12] An oral hearing is also held if the Federal Patent Court considers it expedient[13] or if evidence is to be taken.[14]

[5] Section 73(3) sentence 1 Patent Act.
[6] Section 73(3) sentence 2 Patent Act.
[7] Section 73(3) sentence 3 Patent Act.
[8] Section 76 sentence 1 Patent Act.
[9] Section 76 sentence 2 Patent Act.
[10] Section 77 sentence 1 Patent Act.
[11] Section 77 sentence 2 Patent Act.
[12] Section 78 no. 1 Patent Act.
[13] Section 78 no. 3 Patent Act.
[14] Section 78 no. 2 Patent Act.

5.4 Decision on the Appeal

The appeal proceedings end with a decision.[15] An inadmissible appeal is rejected.[16] An appeal is inadmissible if it is not permitted or was not filed in the statutory form and within the statutory period.[17] In this case, the decision is issued without an oral hearing.[18]

Furthermore, the Federal Patent Court may set aside the contested decision without issuing its own decision.[19] This is done in particular if the DPMA has not yet decided on the merits, if the contested proceedings before the DPMA contain a material error that can be corrected by the DPMA, or if significant new facts have emerged that the DPMA has not yet been able to consider.[20]

If the Federal Patent Court sets aside a decision of the DPMA, it attaches a legal assessment to the decision, which the DPMA must follow.[21]

5.5 Costs of the Appeal Proceedings

An appeal fee must be paid within the appeal period. Otherwise, the appeal is inadmissible and will be rejected without further action.

The Federal Patent Court may order the refund of the appeal fee if this is equitable.[22] The Federal Patent Court may order the DPMA to bear the costs of the proceedings, in particular the costs of the appellant. However, this is only possible if the President of the DPMA was a party to the appeal proceedings and has also filed at least one request.[23]

[15] Section 79(1) Patent Act.
[16] Section 79(2) Patent Act.
[17] Section 79(2) sentence 1 Patent Act.
[18] Section 79(2) sentence 2 Patent Act.
[19] Section 79(3) Patent Act.
[20] Section 79(3) sentence 1 Patent Act.
[21] Section 79(3) sentence 2 Patent Act.
[22] Section 80(3) Patent Act.
[23] Section 80(2) Patent Act.

German Opposition Appeal

6

Contents

The appeal against an opposition decision is the legal remedy against the decision on an opposition. An appeal against an opposition decision is heard before the Federal Patent Court.

6.1 Timeline of an Appeal Against an Opposition Decision

Figure 6.1 shows the chronological sequence of an appeal procedure as a legal remedy against the decision following an opposition procedure. As a rule, the final oral hearing before the Federal Patent Court can be expected approximately two to four years after the oral hearing before the Opposition Division. The timeline begins with the "Oral hearing of the appeal", at which the Patent Office decides on the opposition to a granted patent. This decision by the DPMA is being challenged by an appeal before the Federal Patent Court.

T. H. Meitinger, *Grant of a Patent, Opposition, Appeal, and Invalidity*,
https://doi.org/10.1007/978-3-662-72556-6_6

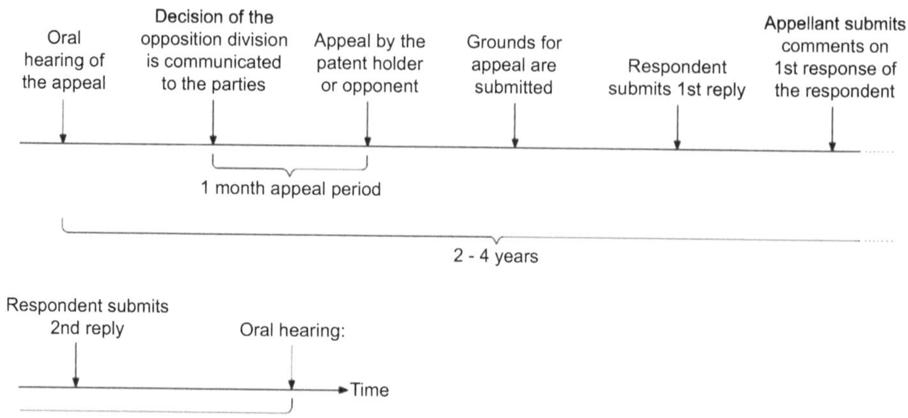

Fig. 6.1 Timeline of an appeal against an opposition decision

6.2 Special Features of the proceedings

The opposition appeal procedure is the judicial instance that can be called upon to review a decision by the Patent Office on an opposition (§ 73 Absatz 1 German Patent Law). It is a contentious procedure in which an opponent and the patent proprietor face each other. The appellant can be either the opponent or the patent proprietor. The appeal must be justified. However, there is no strict deadline for this. Nevertheless, the reasons for the appeal should be submitted promptly. The appeal to the Federal Patent Court is a court of fact, meaning that the court is not limited to reviewing the legal aspects of the Patent Office's decision. Rather, the Patent Office's decision can be reviewed in terms of both fact and law. New facts presented by the parties involved can be taken into account.

7 Utility Model Cancellation Proceedings

Contents

An application for cancellation of a utility model is processed by the Utility Model Division of the German Patent Office.[1] An oral hearing is held only if requested by a party or if the Patent Office considers an oral hearing to be expedient.[2]

7.1 Timeline of a Cancellation Procedure

Figure 7.1 shows the timeline of the utility model cancellation procedure, in which the utility model proprietor must file an objection to defend their intellectual property right. To prepare for the oral hearing, written submissions may be exchanged between the parties prior to the hearing, with the utility model proprietor first responding to the grounds of the application, after which the applicant may submit a statement to the Utility Model Division in response to the proprietor's reply. Finally, the utility model proprietor is given another opportunity to defend the utility model. An oral hearing is held only if requested by one of the parties or if the Utility Model Division considers it expedient.[3]

[1] Section 17 paragraph 3 sentence 1 of the Utility Model Act.

[2] Section 17 paragraph 2 sentence 5 of the Utility Model Act.

[3] Section 17 paragraph 2 sentence 5 of the Utility Model Act.

T. H. Meitinger, *Grant of a Patent, Opposition, Appeal, and Invalidity*,
https://doi.org/10.1007/978-3-662-72556-6_7

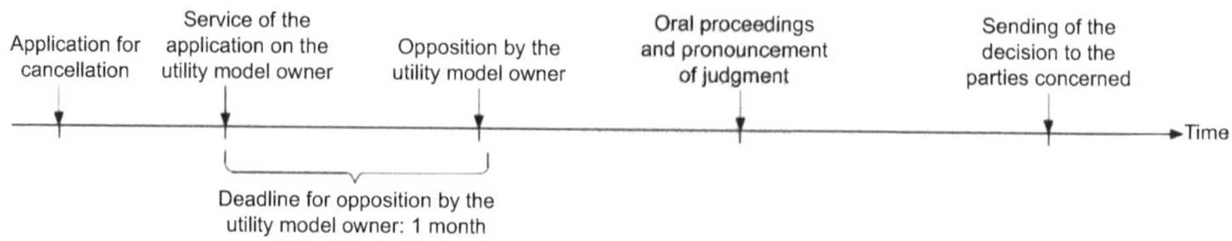

Fig. 7.1 Timeline of a utility model cancellation procedure

7.2 Procedure

A utility model cancellation procedure is initiated upon application. The application must be submitted in writing to the German Patent and Trade Mark Office.[4] In addition, the facts supporting the application for cancellation must be stated.[5]

The German Patent and Trade Mark Office transmits the application for cancellation to the proprietor of the challenged utility model and gives them the opportunity to object within a period of one month.[6] If the proprietor does not object, the utility model is cancelled (default decision).[7]

If the proprietor of the utility model objects to the cancellation, the objection is forwarded to the applicant.[8] The Utility Model Division will schedule an oral hearing if it considers this expedient or if one of the parties requests it.[9]

7.3 Decision of the Utility Model Division

The Utility Model Division decides on the application for cancellation of the utility model by means of a decision.[10] The decision must be reasoned.[11] The decision may be pronounced immediately after the oral hearing.[12]

The decision of the Utility Model Division of the German Patent and Trade Mark Office may be appealed to the Federal Patent Court.[13]

[4] Section 16 sentence 1 of the Utility Model Act.

[5] Section 16 sentence 2 of the Utility Model Act.

[6] Section 17 sentence 1 of the Utility Model Act.

[7] Section 17 sentence 2 of the Utility Model Act.

[8] Section 17 paragraph 2 sentence 1 of the Utility Model Act.

[9] Section 17 paragraph 2 sentence 5 of the Utility Model Act.

[10] Section 17 paragraph 3 sentence 1 of the Utility Model Act.

[11] Section 17 paragraph 3 sentence 2 of the Utility Model Act.

[12] Section 17 paragraph 3 sentence 6 of the Utility Model Act.

[13] Section 18 paragraph 1 of the Utility Model Act.

7.4 Costs of the Procedure

The Utility Model Division decides on the allocation of costs in its decision.[14] A decision on costs is made only upon request if the Utility Model Division does not issue a decision on the merits.[15] The value in dispute is determined by decision of the Utility Model Division upon application by the parties.[16] The value in dispute of the cancellation procedure forms the basis for calculating the costs.

[14] Section 17 paragraph 4 sentence 1 of the Utility Model Act.

[15] Section 17 paragraph 4 sentence 2 of the Utility Model Act.

[16] Section 17 paragraph 5 sentence 1 of the Utility Model Act.

8 Utility Model Appeal Proceedings

Contents

The decisions of the Utility Model Offices and the Utility Model Divisions may be appealed. The appeal is heard before the Federal Patent Court.[1]

8.1 Significance of Utility Model Appeal Proceedings

Figure 8.1 shows the development of the number of utility model appeal cases before the Federal Patent Court. The number of new cases decreased by approximately 24% between 2000 and 2023.

[1] Section 18(1) of the Utility Model Act.

T. H. Meitinger, *Grant of a Patent, Opposition, Appeal, and Invalidity*,
https://doi.org/10.1007/978-3-662-72556-6_8

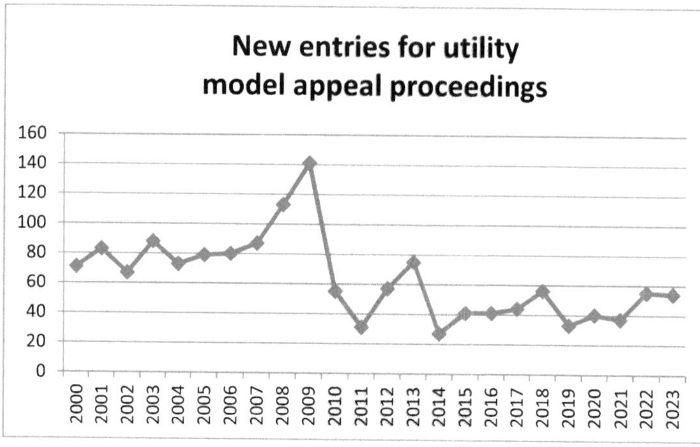

Fig. 8.1 New utility model appeals from 2000 to 2023[2]

8.2 Suspension

If an appeal is pending before the Utility Model Appeal Board against the decision of the Utility Model Division on an application for cancellation of a utility model, the Board may order that parallel proceedings, the outcome of which depends substantially on the existence of the utility model, be suspended until the decision of the Utility Model Appeal Board.[3] If the Appeal Board considers the utility model to be subject to cancellation, it must order the suspension.[4]

8.3 Admission of Legal Appeal

A legal appeal for judicial review of the appeal before the Federal Court of Justice requires the approval of the Appeal Board of the Federal Patent Court.[5]

[2] Federal Patent Court, Annual Reports from 2014 to 2023, https://www.bundespatentgericht.de/DE/Presse/Publikationen/Bilderstrecke_Jahresbericht.html?nn=195898&cms_gcp_225314=0#gcp_anchor_225314, accessed on 21.12.2024. or Library of the Federal Patent Court, Annual Reports from 2007 to 2009, XE 180 ac 1/2000 to XE 180 ac 1/2009.

[3] Section 19 sentence 1 of the Utility Model Act.

[4] Section 19 sentence 2 of the Utility Model Act.

[5] Section 18(4) sentence 1 of the Utility Model Act.

Legal Appeal Before the Federal Court of Justice (BGH)

9

Contents

Appeals against the decisions of the Boards of Appeal of the Federal Patent Court are heard by the Federal Court of Justice. Such an appeal is heard before the Xth Chamber of the Federal Court of Justice.

An appeal on points of law is only admissible in rare exceptional cases. In particular, such an appeal is possible if it is admitted or if the proceedings or the decision of the Federal Patent Court contain serious errors.

9.1 Timeline of an Appeal on Points of Law

Figure 9.1 shows the timeline of an appeal on points of law, whereby an appeal may be filed within one month from the service of the contested decision of the Federal Patent Court.[1]

[1] Section 102(1) Patent Act.

T. H. Meitinger, *Grant of a Patent, Opposition, Appeal, and Invalidity*,
https://doi.org/10.1007/978-3-662-72556-6_9

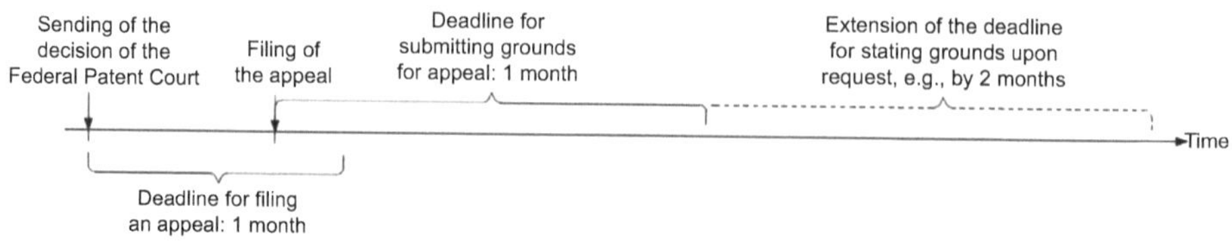

Fig. 9.1 Timeline of an appeal on points of law

The appeal on points of law must be substantiated.[2] The substantiation must be submitted to the Federal Court of Justice within one month after filing the appeal.[3] The one-month period for preparing the substantiation of an appeal on points of law is short. It is therefore advisable to apply to the presiding judge of the competent chamber for an extension of the deadline, for example by two months.[4] It can be assumed that the extension will be granted.

9.2 Admission of the Appeal on Points of Law

The Federal Patent Court must admit an appeal on points of law if a fundamental legal question needs to be clarified, which would enable consistent future case law.[5] Furthermore, an appeal on points of law must be admitted if legislative gaps are to be closed by judicial development of the law.[6]

9.3 Appeal on Points of Law without Admission

An appeal on points of law that does not require admission is particularly admissible if the principle of the right to be heard has been violated, i.e., if the appellant was not able to comment on circumstances relevant to the decision.[7] The second significant reason for an appeal on points of law without admission is if the contested decision of the Federal Patent Court does not contain reasons.[8]

[2] Section 102(3) sentence 1 Patent Act.
[3] Section 102(3) sentence 2 Patent Act.
[4] Section 102(3) sentence 2 Patent Act.
[5] Section 100(2) no. 1 Patent Act.
[6] Section 100(2) no. 2 Patent Act.
[7] Section 100(3) no. 3 Patent Act.
[8] Section 100(3) no. 6 Patent Act.

9.4 Legal Violation

Only a party to the appeal proceedings may file an appeal on points of law.[9] An appeal on points of law may only be used to bring a legal violation before the Federal Court of Justice for adjudication.[10]

9.5 Substantiation of the Appeal on Points of Law

The substantiation of the appeal on points of law must explain to what extent the decision of the Federal Patent Court is being challenged.[11] The substantiation must state the legal provision that is allegedly violated by the lower court.[12] If the appeal on points of law challenges the appeal proceedings before the Federal Patent Court, the facts giving rise to the legal violation must be specified.[13]

9.6 Proceedings

The Federal Court of Justice will first examine the admissibility of the appeal on points of law.[14] If the appeal on points of law was not filed and substantiated within the statutory deadlines, or if it was not admitted or is not admissible without admission, the appeal will be dismissed as inadmissible.[15]

If the appeal on points of law is admissible, the statement of appeal will be forwarded to the other parties. The other parties are also given the opportunity to submit statements within a specified period.[16]

If the contested decision of the Federal Patent Court is set aside and remitted to the Federal Patent Court, the Federal Patent Court is bound by the legal assessment of the Federal Court of Justice.[17]

---

[9] Section 101(1) Patent Act.
[10] Section 101(2) sentence 1 Patent Act.
[11] Section 102(4) no. 1 Patent Act.
[12] Section 102(4) no. 2 Patent Act.
[13] Section 102(4) no. 3 Patent Act.
[14] Section 104 sentence 1 Patent Act.
[15] Section 104 sentence 2 Patent Act.
[16] Section 105(1) sentence 1 Patent Act.
[17] Section 108 Patent Act.

9.7 Costs of the Appeal on Points of Law

The costs of the proceedings may be imposed on a party if this is equitable. However, this only applies to those costs that were necessary due to the proceedings.[18] If the appeal on points of law is inadmissible or is dismissed, the appellant must bear the costs.[19] If costs arise due to gross fault on the part of a party, those costs are imposed on that party.[20] If the President of the German Patent and Trade Mark Office was a party, costs may only be imposed on him if he filed the appeal on points of law or submitted requests.[21]

9.8 Admitted Attorney-at-Law

In proceedings before the Federal Court of Justice, the parties must be represented by an attorney-at-law admitted to practice before the Federal Court of Justice, known as a BGH attorney.

[18] Section 109(1) sentence 1 Patent Act.

[19] Section 109(1) sentence 2 Patent Act.

[20] Section 109(1) sentence 3 Patent Act.

[21] Section 109(2) Patent Act.

Nullity Proceedings 10

Contents

A nullity action can be used to challenge a German patent or a European patent effective in Germany. An action for a declaration of nullity of a patent can be brought before the Federal Patent Court.[1] The action must be directed against the patent proprietor registered in the register.[2]

The nullity proceedings are adversarial proceedings in which a nullity plaintiff and a patent proprietor face each other. The nullity action is a formative action, in which the contested patent can be revoked or maintained in an amended form. If the nullity action is unsuccessful, the action is dismissed.

In first-instance nullity proceedings before the Federal Patent Court, there is no requirement to be represented by a lawyer. In contrast, representation by a lawyer is mandatory in nullity appeal proceedings before the Federal Court of Justice. However, given the complexity of the proceedings, representation by a patent attorney or an

[1] Section 81(1) sentence 1 of the Patent Act.
[2] Section 81(1) sentence 2 of the Patent Act.

T. H. Meitinger, *Grant of a Patent, Opposition, Appeal, and Invalidity*,
https://doi.org/10.1007/978-3-662-72556-6_10

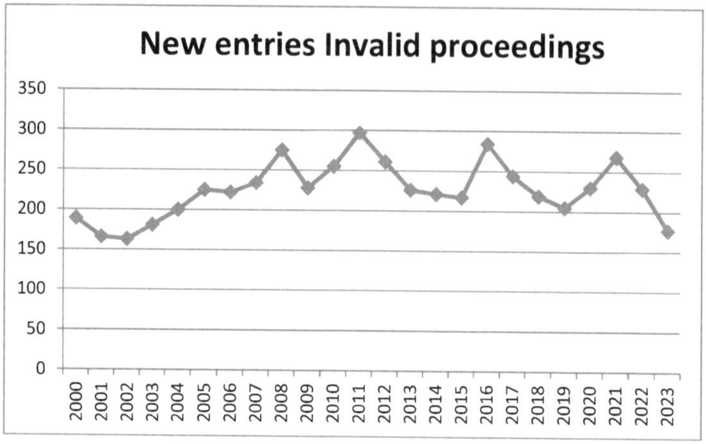

Fig. 10.1 New nullity actions filed from 2000 to 2023[3]

experienced lawyer is recommended. For legally and technically demanding disputes, it may even be advisable to be represented by both a patent attorney and a lawyer.

10.1 Significance of Nullity Proceedings

Figure 10.1 shows the annual number of new nullity actions filed. The illustration shows that the number of nullity cases is relatively low, fluctuating between 300 and 150 cases. In 2023, only 177 new applications for a declaration of nullity were filed with the Federal Patent Court. Between 2000 and 2023, the number of nullity actions filed annually decreased by approximately 6%.

10.2 Timeline of Nullity Proceedings

Figure 10.2 shows the timeline of nullity proceedings, with the illustration being abbreviated, as the nullity plaintiff is still given an opportunity to comment on the grounds for the patent proprietor's opposition, and the patent proprietor is given an opportunity

[3]Federal Patent Court, Annual Reports from 2014 to 2023, https://www.bundespatentgericht.de/DE/Presse/Publikationen/Bilderstrecke_Jahresbericht.html?nn=195898&cms_gcp_225314=0#gcp_anchor_225314, accessed on 21.12.2024. or Library of the Federal Patent Court, Annual Reports from 2000 to 2009, XE 180 ac 1/2000 to XE 180 ac 1/2009.

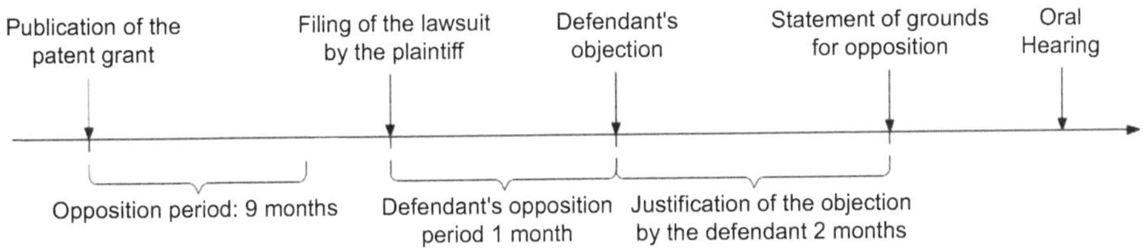

Fig. 10.2 Timeline of a nullity proceeding

to respond to the plaintiff's submission, in order to thoroughly prepare the oral hearing before the Nullity Senate of the Federal Patent Court.

10.3 Admissibility

Given that a nullity action can be filed at any time, defects in admissibility can be remedied by subsequent submission. However, defects in admissibility can lead to undesirable delays in the proceedings.

The action is inadmissible if the period for filing an opposition has not yet expired or as long as opposition proceedings are still pending.[4]

The action can be brought by any person. No legal interest is required. However, in the case of the ground of unlawful removal, only the injured party may bring an action.[5]

The action must be filed in writing.[6] The statement of claim and the documents relating to the action are transmitted by the Federal Patent Court to the patent proprietor.[7]

The statement of claim must specify the plaintiff, the defendant, and the subject matter of the dispute. The subject matter of the dispute includes the claim (complete or partial declaration of nullity of the patent) and the grounds for the claim (for example, lack of novelty, lack of inventive step, insufficient disclosure, or inadmissible extension).[8]

The statement of claim must specify the facts and evidence on which the nullity action is based.[9] The facts and evidence should be attached to the statement of claim and described therein in such a way that they can be readily understood by a person skilled in the art.

[4] Section 81(2) sentence 1 of the Patent Act.
[5] Section 81(3) of the Patent Act.
[6] Section 81(4) sentence 1 of the Patent Act.
[7] Section 81(4) sentence 3 of the Patent Act.
[8] Section 81(5) sentence 1 of the Patent Act.
[9] Section 81(5) sentence 2 of the Patent Act.

10.4 Grounds for Nullity

Nullity proceedings can be used to judicially review the legal validity of a German patent and a European patent validated for Germany. The grounds for nullity of a German patent are set out in Section 22 in conjunction with Section 21 of the German Patent Act. For a European patent validated in Germany, Article II Section 6 of the IntPatÜG (Gesetz über internationale Patentübereinkommen) applies.

Lack of Patentability

The ground of lack of patentability applies if the provisions of Sections 1 to 5 of the Patent Act are met. Lack of patentability exists if:

- **no technical character is present,**[10] Patent law is only accessible to technical inventions, whereby the legislator has not defined what constitutes an "invention" under the Patent Act. This gap was deliberately left by the legislator to the legal practice of the Patent Office and, above all, to the case law of the Federal Patent Court and the Federal Court of Justice. This allows for a continuous adaptation of the concept of invention to the current understanding of technology.
- **if it concerns discoveries, scientific theories, or mathematical methods,**[11] Patent law does not serve to promote scientific research, but rather the development of technology.
- **an aesthetic design without technical function is present,**[12] Patent law is not accessible to aesthetic creations of form. For aesthetic designs, protection can be sought under the Design Act.
- **plans, rules, and methods for mental activities, for games, or for business activities (for example, a business model) or computer programs,**[13] In particular, business models cannot be patented. Only computer programs "as such" are excluded from patent protection. If a computer program exhibits technical features, it is nevertheless patentable.
- the mere presentation of information,[14]
- **invention violates public order or morality,**[15]
- **lack of novelty,**[16]

[10] Section 21 paragraph 1 no. 1 in conjunction with Section 1 paragraph 1 Patent Act.
[11] Section 21 paragraph 1 no. 1 in conjunction with Section 1 paragraph 3 no. 1 Patent Act.
[12] Section 21 paragraph 1 no. 1 in conjunction with Section 1 paragraph 3 no. 2 Patent Act.
[13] Section 21 paragraph 1 no. 1 in conjunction with Section 1 paragraph 3 no. 3 Patent Act.
[14] Section 21 paragraph 1 no. 1 in conjunction with Section 1 paragraph 3 no. 4 Patent Act.
[15] Section 21 paragraph 1 no. 1 in conjunction with Section 2 paragraph 1 Patent Act.
[16] Section 21 paragraph 1 no. 1 in conjunction with Section 3 paragraph 1 sentence 1 Patent Act.

- **no inventive step**[17] or
- **no industrial applicability is present,**[18] The ground for revocation of lack of industrial applicability plays no practical role, as there is hardly any product that is not marketable.

In addition, grounds for action are:

- **no clear and complete disclosure of the invention in the patent,**[19] An invention is executable if a person skilled in the art can carry out the invention according to the description without inventive activity. The skilled person is expected to perform a creative act and, to a certain extent, experiments, but without reaching the level of an inventive step as defined by the Patent Act. The skilled person must therefore be able to realize the invention based on their expertise and skills, and must not have to overcome excessive difficulties to achieve the inventive effect.[20] It is not necessary for all features required to realize the invention to be included in the patent claim. It is sufficient if the skilled person can carry out the invention from the claim using the description in the patent specification.[21] In particular, it is sufficient to indicate the direction to the skilled person in order to arrive at the invention.[22] It is not necessary to describe the current state of the art in the patent application.[23]
- **Impermissible amendment of the original disclosure,**[24] the ground for nullity exists if the disclosure of the patent goes beyond what was filed with the Patent Office on the filing date.
- **Extension of the scope of protection of the patent**[25] exists if the current disclosure of the patent specification goes beyond a previously patented version. The ground for nullity is present if an embodiment according to the wording of a patent claim was not included in a previous version of the patent.[26] If features are combined in a patent claim that are not related to each other in the patent specification, this may already constitute an impermissible extension.

---

17 Section 21 paragraph 1 no. 1 in conjunction with Section 4 sentence 1 Patent Act.

18 Section 21 paragraph 1 no. 1 in conjunction with Section 5 Patent Act.

19 Section 21 paragraph 1 no. 2.

20 BGH GRUR 2010, 901 polymerizable cement mixture; BGH GRUR 1984, 272 insulating glass pane edge joint filling device; BGH Mitt. 2002, 176 intercom system; BGH GRUR 1980, 166, 168 dual axle assembly; BGH GRUR 2010, 916 stapling device.

21 BGH GRUR 1998, 899, 900 alpine ski; BGH GRUR 2003, 223, 225 coupling device II.

22 BGH GRUR 1979, 461 color picture tube.

23 BGH GRUR 1984, 272 insulating glass pane edge joint filling device.

24 Section 21 paragraph 1 no. 4 first half-sentence Patent Act.

25 Section 22 paragraph 1 Patent Act.

26 BGH GRUR 2010, 1084 wind energy converter.

- **Unlawful removal,** in the case of unlawful removal, it is not the patentability but the entitlement of the patent holder that must be examined.

The following possible deficiencies **cannot** constitute grounds for nullity action:

- erroneous claiming of a priority
- defects in the grant procedure[27]
- lack of unity[28]
- failure to solve the problem,
- an embodiment in the description that is not covered by the scope of protection,
- the patent proprietor has no legitimate interest in legal protection for the patent,[29]
- product-by-process claim, although another claim format would be possible.[30] A product-by-process claim describes a device by its manufacturing process instead of listing the individual components of the device.
- lack of clarity,
- double patenting,[31]
- incorrect designation of the inventor and
- fraudulent obtaining of a patent.[32] Fraudulent obtaining of a patent arises, for example, from false statements during the patent grant procedure, if the patent is granted as a result of these statements.

10.5 Suspension

The separation principle in Germany states that proceedings for infringement of a patent and nullity proceedings against the same patent are conducted in two different proceedings before different courts.[33]

In Germany, the infringement court[34] generally cannot question the validity of a granted patent.[35] The validity of a patent cannot be examined in infringement proceedings,

[27] BGH GRUR 1965, 473 Dauerwellen I.

[28] BGH Liedl 1967/1968 Warenzufuhrvorrichtung.

[29] BGH GRUR 1991, 376 bullet-resistant metal door.

[30] BGH GRUR 1997, 612 polyethylene filaments.

[31] BGH GRUR 1991, 376 bullet-resistant metal door.

[32] BGH GRUR 1954, 107 multiple clamp.

[33] Pagenberg GRUR Int. 2010, 195.

[34] § 143 Patent Act.

[35] BGH GRUR 1979, 624 convertible shooting target; BGH GRUR 2002, 550 judge disqualification; BGH GRUR 2004, 710 printing machine temperature control system I.

but only in opposition or nullity proceedings. Therefore, in infringement proceedings, the patent as granted must be accepted.[36]

To avoid having to decide on an infringement while a promising opposition or nullity proceeding is pending, the infringement proceedings may be suspended until the end of the opposition or nullity proceedings.[37]

Suspension serves to avoid contradictory decisions. Therefore, if the decision in a first proceeding significantly affects the course of a second proceeding, the second proceeding may be suspended until a decision is reached in the first. However, if it is assumed that the first proceeding will very likely be decided in such a way that the second proceeding will not be affected in its basis, suspension usually does not take place. Suspension of a patent infringement proceeding is only considered if an opposition or nullity proceeding is pending.

If a foreign part of a European patent is challenged, this does not constitute grounds for suspending infringement proceedings based on the German part of the European patent.[38]

Suspension of infringement proceedings does not occur automatically if an opposition or nullity action is filed. In addition, there must be a sufficient likelihood of success for the opposition or nullity proceedings.[39] Mere doubts about validity are not sufficient to suspend infringement proceedings. However, at the latest, the declaration of nullity by the Federal Patent Court justifies suspension, even if the judgment is still subject to appeal.[40] This does not apply, however, if the decision of the Federal Patent Court is clearly erroneous.[41]

10.6 Distinction between nullity proceedings and opposition proceedings

The grounds for nullity essentially correspond to the grounds for opposition. Revocation in opposition proceedings and a declaration of nullity are effective against everyone, so that a revoked patent does not confer any right to prohibit anyone.

[36] BGH GRUR 1959, 320 moped clutch; BGH GRUR 1964, 606, 609 conveyor belt; BGH GRUR 1970, 296 multipurpose agricultural machine; BGH GRUR 1979, 624 f. convertible shooting target.

[37] R. Rogge GRUR Int. 1996, 386, 390; BGH GRUR 2004, 710 printing machine temperature control system I.

[38] OLG Düsseldorf 29.06.2000 2 U 76/99.

[39] BGH GRUR 1959, 320, 324 moped clutch; OLG Düsseldorf Mitt. 1997, 257; OLG Düsseldorf 14.11.1996 2 U 28/93; OLG Düsseldorf 22.12.2008 2 U 65/07; OLG Düsseldorf 14.06.2007 2 U 135/05 GRUR-RR 2008, 333.

[40] Maltzahn GRUR 1985, 172.

[41] OLG Düsseldorf Mitt. 2008, 327 olanzapine; Müller-Stroy/Bublak/Coehn Mitt. 2008, 335, 336.

- Deadlines: For an opposition, the 9-month period after publication of the grant of the patent must be observed. A nullity action is not subject to a time limit and can be filed at any time after the opposition period has expired or any pending opposition proceedings have been concluded.
- The opposition procedure is an administrative procedure subsequent to and related to the patent grant procedure. In contrast, nullity proceedings are adversarial judicial proceedings that are concluded by a judgment.
- In first-instance opposition proceedings, the opposition division is not bound by the requests of the opponent. The opposition division may examine grounds for revocation in addition to those asserted by the opponent.[42] In nullity proceedings, the nullity senate of the Federal Patent Court is bound by the grounds for nullity put forward by the nullity plaintiff.[43]
- Neither opposition nor opposition appeal proceedings have a binding effect on subsequent nullity proceedings. An opposition procedure does not prejudice a subsequent nullity procedure.[44] This also applies if no new facts or evidence are presented.[45] It is therefore quite possible that in nullity proceedings a patent is revoked on the basis of prior art that was not considered relevant in a previous opposition procedure.[46] However, in nullity proceedings, a previous opposition or opposition appeal procedure is regarded as a relevant legal opinion, and its assessment of the facts and evidence is taken seriously.

10.7 Example

Excerpts from an actual nullity pleading are presented and discussed.

[42] BGH GRUR 1995, 333 aluminum trihydroxide; BPatG GRUR 1986, 605; BPatG GRUR 1991, 40; BPatG GRUR 2004, 356.

[43] BPatG GRUR Int. 1996, 822.

[44] BGH GRUR 2005, 967 radiation control.

[45] BGH GRUR 2005, 967 radiation control.

[46] BGH GRUR 1996, 757 gear cutter; NGH GRUR 1996, 862, 864 arc segment; BGH GRUR 1998, 895 stormwater basin.

The subject of a nullity action can be formulated as follows:

File number: y Ni xx

Action for annulment of the German part (DE 501 xx) of European patent EP xx

Plaintiff: xx oHG

Defendant: yy KG

Our file: G0002/TM

On behalf of and acting for our client,

xx oHG

an action for annulment is brought against European patent EP xx

nullity action

is being brought.

1. Feature Analysis of Claim 1 of the Patent in Suit
The following feature analysis is used as the basis for discussing the subject matter of claim 1:

A feature analysis follows, which should be included separately in the appendix. Subsequently, the subject matter of the main claim or the subject matters of the independent claims can be interpreted in light of the description.

Afterwards, the individual documents of the prior art are listed and their disclosure briefly explained.

In the statement of grounds for the action, it must be set out why the independent claims meet the grounds for revocation and are therefore not novel, not inventive, not executable by a person skilled in the art, and/or have been impermissibly extended.

2. Lack of Novelty over DE xx (NK3a)
The published application DE xx describes an embodiment of a tool according to Figs....

Therefore, from the embodiment of Figs. 14, 15, 16, and 17 of NK3a, all features of the subject matter of claim 1 are disclosed. Claim 1 is therefore anticipated by NK3a and lacks novelty.

It must be described where in the prior art document the individual features of the main claim are disclosed, thereby demonstrating that the main claim is anticipated by this document and lacks novelty.

The further prior art documents that call into question novelty and inventive step are analyzed. If there is a lack of enablement, this must be set out. Impermissible extension is examined on the basis of the originally filed application documents.

If the patent in suit claims priority from an earlier application, it should be examined to what extent the claim to priority is justified. For the dependent claims, it must be set out in the same way why their features do not lead to a legally valid claim.

3. Summary

In accordance with the arguments set out above, the subject matter of claim 1 of the patent in suit is not novel, or at least not inventive, and as set out in the nullity action, the same result can be assumed for the dependent claims of the patent in suit. The claims of the patent in suit are therefore not patentable.

The patent in suit is therefore to be declared null and void.

11 Nullity Appeal

Contents

The judgments of the nullity senates can be challenged by means of an appeal. The nullity appeal is heard before the Federal Court of Justice.[1] The appeal proceedings are initiated by filing the notice of appeal with the Federal Court of Justice.[2]

11.1 Timeline of the Nullity Appeal

Figure 11.1 shows the timeline of a nullity appeal, which can be lodged with the Federal Court of Justice within one month after service of the full judgment. However, the appeal must be filed no later than five months after pronouncement of the judgment if no service of the judgment has taken place.[3] The grounds for appeal may be submitted together with the notice of appeal or subsequently within three months after service of

[1] Section 110(1) Patent Act.
[2] Section 110(2) Patent Act.
[3] Section 110(3) Patent Act.

T. H. Meitinger, *Grant of a Patent, Opposition, Appeal, and Invalidity*,
https://doi.org/10.1007/978-3-662-72556-6_11

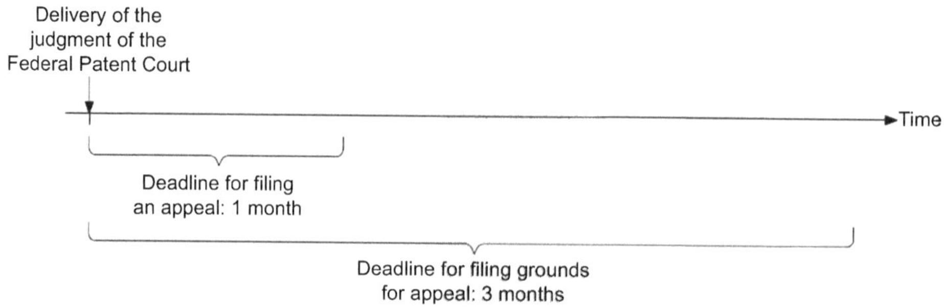

Fig. 11.1 Timeline of a nullity appeal

the contested judgment.[4] Upon request, the deadline for submitting the grounds may be extended if the opposing party consents.[5]

11.2 Deadline and Form

The appeal must be filed within one month after service of the judgment or, if no service has taken place, no later than five months after pronouncement of the judgment.[6]

The notice of appeal must specify the judgment being challenged.[7] In addition, the notice of appeal must include a statement that an appeal is being lodged against this judgment.[8]

An appeal is only admissible if it is based on a legal error, meaning that a legal provision was incorrectly applied or not applied at all by the Federal Patent Court.[9]

A legal error exists in particular if the Patent Court was not properly constituted, if there are grounds to suspect bias on the part of a judge, or if the decision of the Federal Patent Court lacks reasoning.[10]

[4] Section 112(2) sentences 1 to 3 Patent Act.
[5] Section 112(2) sentence 4 Patent Act.
[6] Section 110(3) Patent Act.
[7] Section 110(4) no. 1 Patent Act.
[8] Section 110(4) no. 2 Patent Act.
[9] Section 111(2) Patent Act.
[10] Section 111(3) nos. 1, 3, and 6 Patent Act.

11.3 Statement of Grounds

The nullity appeal must be substantiated.[11] The statement of grounds for appeal must indicate the extent to which the judgment is being challenged.[12] Furthermore, the facts and circumstances from which the legal error arises must be specified.[13]

11.4 Proceedings

The Federal Court of Justice will first examine whether a legal error has been asserted and whether the appeal was filed and substantiated in due time.[14] If these requirements are not met, the appeal will be dismissed as inadmissible.[15]

If the appeal is admissible, a date for the oral hearing will be set and communicated to the parties.[16]

The Federal Court of Justice will only examine the matter to the extent requested by the parties.[17]

An amendment of the claim and a defense of the patent in an amended form are only permissible if the opposing party consents or if the Federal Court of Justice considers it expedient.[18]

An amendment of the claim occurs when the claim is changed, thereby altering the subject matter of the dispute. For example, an amendment would be if the action no longer challenges the main claim but instead a dependent claim of a patent.[19]

---

11 Section 112(1) Patent Act.

12 Section 112(3) no. 1 Patent Act.

13 Section 112(3) no. 1 Patent Act.

14 Section 114(1) sentence 1 Patent Act.

15 Section 114(1) sentence 2 Patent Act.

16 Section 114(3) Patent Act.

17 Section 116(1) Patent Act.

18 Section 116(2) Patent Act.

19 Section 263 Code of Civil Procedure (ZPO).

11.5 Decision of the Federal Court of Justice

The Federal Court of Justice renders its judgment after an oral hearing.[20] No oral hearing is held if the parties consent or if only a decision on costs is required.[21] If the parties do not appear at the oral hearing, the decision is made on the basis of the record. If only one party fails to appear, the hearing proceeds without that party.[22]

If the legal error can be confirmed, but the contested decision of the Federal Patent Court nevertheless proves to be correct for other reasons, the appeal will be dismissed.[23]

If the appeal is well-founded, the contested decision will be set aside to that extent[24] and remitted to the Federal Patent Court.[25] It may be remitted to a different nullity senate.[26] In its decision, the Federal Patent Court must take into account the legal assessment of the Federal Court of Justice.[27]

If the case is ready for decision, the Federal Court of Justice will decide itself and not remit the case.[28]

11.6 Mandatory Representation

The parties must be represented by an attorney-at-law or a patent attorney.[29]

---

[20] Section 118(1) sentence 1 Patent Act.

[21] Section 118(3) Patent Act.

[22] Section 118(4) Patent Act.

[23] Section 119(1) Patent Act.

[24] Section 119(2) sentence 1 Patent Act.

[25] Section 119(3) sentence 1 Patent Act.

[26] Section 119(3) sentence 2 Patent Act.

[27] Section 119(4) Patent Act.

[28] Section 119(5) Patent Act.

[29] Section 113 sentence 1 Patent Act.

European Opposition Procedure 12

Contents

The European opposition procedure is similar to the German procedure. In particular, the grounds for revocation are identical. Nevertheless, there are some differences between the European opposition procedure and the German procedure, which will be addressed in detail.

A European patent or a unitary patent can be challenged by means of an opposition. The competent authority for the opposition procedure is an Opposition Division of the European Patent Office.[1]

12.1 Significance of the European Opposition Procedure

The number of oppositions before the EPO shows a slightly declining trend, with the number of oppositions decreasing by approximately 16% from 2000 to 2013 (see Fig. 12.1).

[1] Article 19(1) of the European Patent Convention (EPC).

T. H. Meitinger, *Grant of a Patent, Opposition, Appeal, and Invalidity*,
https://doi.org/10.1007/978-3-662-72556-6_12

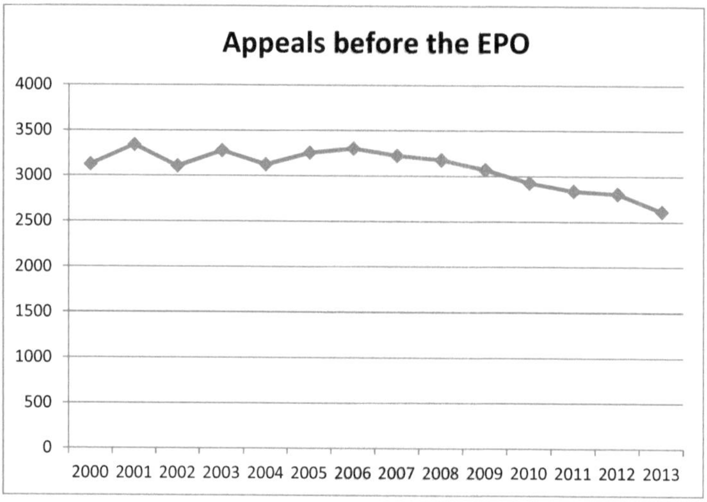

Fig. 12.1 Number of oppositions before the EPO from 2000 to 2013[2]

12.2 Timeline of a European Opposition Procedure

Figure 12.2 shows a chronological sequence of an opposition procedure before the European Patent Office. An opposition can be filed within a period of nine months after the publication of the notice of grant of the European patent.

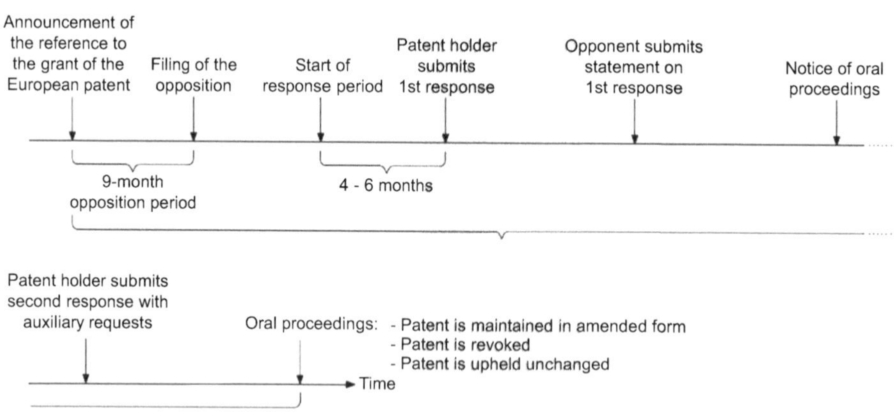

Fig. 12.2 Timeline of a European opposition procedure

[2] European Patent Office, Register, https://register.epo.org/advancedSearch?lng=de&clnrefer=yes, accessed on 21.05.2025.

12.3 Example of an Opposition Brief

An opposition brief from an opponent and a response from a patent proprietor to an opposition are presented and explained.

The contested patent EP 3 516 011 B1 protects a process for pyrolysis, which can be used to produce biochar. In pyrolysis, a thermochemical conversion of organic starting material takes place by heating in the absence of oxygen to produce biochar or charcoal.

The contested patent describes a plant for carrying out the patented process.

Figure 12.3 shows a plant according to the technical teaching of the contested patent with retorts 1, in which the pyrolysis process takes place. The pyrolysis gases generated by the pyrolysis process are collected in a cavity 30 around the retort 1 and fed via a line 41 to a combustion chamber 4. In the combustion chamber 4, the pyrolysis gases are ignited and the resulting hot flue gases are returned via a line 44 to the retort 1, so that the hot flue gases can heat the biogenic starting material 2, thereby initiating or maintaining the pyrolysis process.

The opposition brief is explained below:[3]

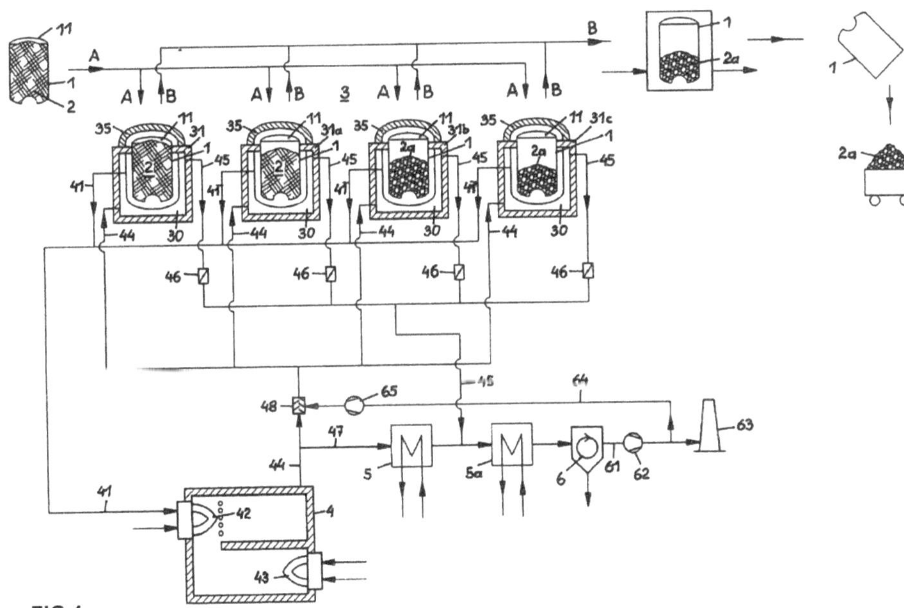

FIG.1

Fig. 12.3 Fig. 1 of EP3516011B1 (DPMA, https://depatisnet.dpma.de/DepatisNet/depatisnet?action=pdf&docid=EP000003516011B1&xxxfull=1, accessed on 21.5.2025)

[3] European Patent Office, https://register.epo.org/application?documentId=E6JTN9735375DSU&number=EP17769097&lng=de&npl=false, accessed on 09.03.2025.

Claus Egger v. Leo Schirrnhofer

Opposition to European Patent No. 3 516 011 B1

On behalf of and in the name of our client, Mr.

Claus Egger
Salmannskirchen 8
85461 Bockhorn

an opposition is hereby filed against European Patent EP 3 516 011 B1

is hereby lodged

is hereby lodged.

Requests

It is requested that patent EP 3 516 011 B1—hereinafter referred to as the contested patent—be revoked in its entirety.

In the alternative, and in the event that the Opposition Division reaches a conclusion that precludes opposition to the contested patent in its entirety, a request is made for oral proceedings.

The grounds for opposition are that the subject matter of the patent is not patentable. In particular, the valid claims of the contested patent are prejudiced by the prior art and are not based on an inventive step. Furthermore, the invention is not disclosed in a manner that is sufficiently clear and complete for a person skilled in the art to carry it out. In addition, the main claim has been extended in an inadmissible manner.

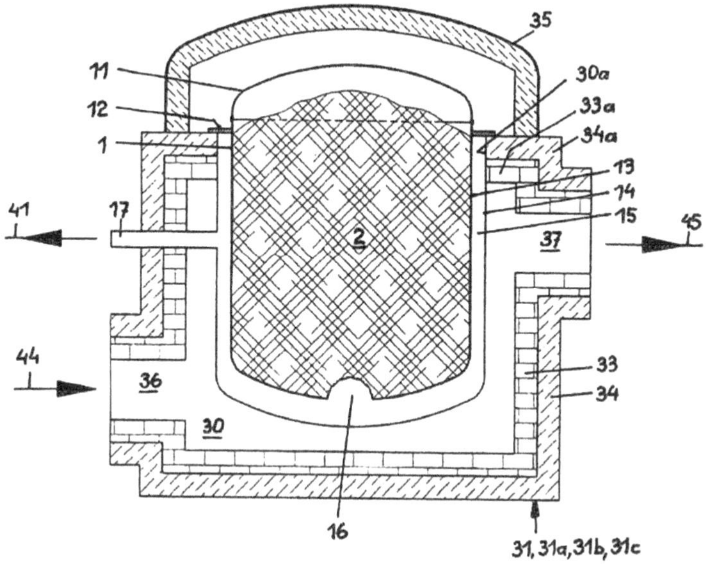

Fig. 12.4 Fig. 1A of EP3516011B1 (DPMA, https://depatisnet.dpma.de/DepatisNet/depatisnet?action=pdf&docid=EP000003516011B1&xxxfull=1, accessed on 21.5.2025)

This opposition brief raises the most commonly asserted grounds for revocation, namely lack of novelty, lack of inventive step, insufficient disclosure for the skilled person, and inadmissible extension.[4]

"Reasons

1. Contested patent

Claim 1 of the contested patent relates to a method for producing biochar."

At this point, it would be advantageous, for example by means of the drawings of the contested patent, to convey one's own interpretation of the technical teaching of the contested patent and thus create understanding for one's own perspective. For example, the following could be included:

Figure 12.4 shows a reactor chamber 31 in which the retort 1 is arranged. The retort 1 contains the biogenic starting material, in particular wood chips, which are converted into charcoal by heating in the absence of oxygen. During this pyrolysis process, pyrolysis gases are generated in the biogenic starting material, which can pass through an

[4] European Patent Office, https://register.epo.org/application?documentId=E6JTN9735375DSU&number=EP17769097&lng=de&npl=false, accessed on 09.03.2025.

opening 16 into the annular space 15 formed between the outer wall 13 of the retort 1 and the partition wall 14. The pyrolysis gases are combusted, and the resulting hot flue gases are introduced into the reactor chamber 30, heating the biogenic starting material 2 and thereby maintaining the pyrolysis process.

> *"For better clarity, the following feature analysis of claim 1 is used as a basis in the following considerations, which is attached in a separate version for easier handling."*[5]

The first step is to break down the claims, at least the independent claims, into their individual features.

"The features of claim 1 of the contested patent can be structured as follows:

M1) Method for producing biochar,
M2) in which biogenic starting material (2) located in retorts (1) is pyrolyzed and
M3) the combustible pyrolysis gases produced by pyrolysis
M4) are combusted in a combustion chamber (4) to generate hot flue gases, wherein
M5) the retorts (1) are introduced sequentially into at least one reactor chamber (31, 31a, 31b, 31c) and
M6) in these, the pyrolysis is carried out by means of the flue gases directed into the at least one reactor chamber (31, 31a, 31b, 31c), wherein
M7) the retorts (1) are at least largely sealed against the entry of hot flue gases and
M8) the heating of the starting materials (2) located in the retorts (1)
M9) is effected by means of the flue gases
M10) by heating the retorts (1)
M11) only indirectly through a partition wall (14), and wherein
M12) a space surrounding the respective retort (1) is an annular space (15)
M13) between the partition wall (14) and an outer wall (13) of the retort, wherein
(according to the preamble)
M14) the pyrolysis gases are conducted through the annular space (15)
M15) to the combustion chamber (4), and that
M16) the flue gases heat the outgoing pyrolysis gases and
M17) the outer wall (13) of the retort (1).
(according to the characterizing part)

2.1 Subject matter of claim 1

The subject matter of the main claim is a method for pyrolysis, i.e., the thermal conversion of organic starting material into pyrolysis gases and biochar in the absence of oxygen.
Such pyrolysis methods are known in the prior art. It is also known to produce biochar by a continuous pyrolysis process in which the biogenic starting material is exposed to hot flue gases. It is also known to combust the pyrolysis gases to generate flue gases that serve to heat the reactor in which the pyrolysis takes place.
A disadvantage of using flue gases to treat the biogenic starting materials is that the flue gases absorb pollutants that arise during pyrolysis.
The technical teaching of the contested patent therefore sets itself the task of ensuring a separation of the flue gases from the pyrolysis process, while still enabling the heating of the biogenic starting materials to induce pyrolysis by means of the flue gases.

[5] European Patent Office, https://register.epo.org/application?documentId=E6JTN9735375DSU&number=EP17769097&lng=de&npl=false, accessed on 09.03.2025.

This object is achieved in that the retorts in which the pyrolysis takes place are so isolated that the flue gases cannot enter the retorts. Heating of the biogenic starting materials located in the retorts by the flue gases therefore takes place through partition walls."[6]

In this section of the statement of opposition, the objective of the technical teaching of the contested patent is elaborated. It is advantageous here to establish that the documents of the prior art serve a similar purpose and therefore represent relevant prior art.

"2.2 Prior Art

In the present opposition, the following documents are newly introduced:

E1 Charcoal Production Processes: an Overview

E2 Charcoal Production with reduced Emissions

E3 Historical Developments of Pyrolysis Reactors: A Review

E4 Technical book "Technologie der Holzverkohlung" by Max Klar from the year 1910

The technical book "Technologie der Holzverkohlung" describes the fundamental aspects of wood carbonization and constitutes prior art prejudicial to novelty.

The following documents are cited in the international search report prepared for the patent application underlying the contested patent:

D1 US 5 725 738 A

D2 CH 237 758 A

D3 DE 10 2014 015815

Copies of the publications E1, E2, and E3 are attached as annexes."[7]

For each document of the prior art, the filing or priority date should always be indicated. For articles from scientific journals, the day or month of publication should be specified. This demonstrates that the respective documents are relevant, as they have an earlier effective date compared to the contested patent.

"2.3 Lack of Novelty

2.3.1 Lack of Novelty due to "Charcoal Production Processes: an Overview" (E1)

E1 deals with the production of biochar ("Charcoal production processes", title, feature M1). Here, a "biomass pyrolysis" is described, i.e., the pyrolysis of a biogenic material (page 19, left column, second sentence from the top; feature M2).

On page 23, left column, third paragraph, it is described that in retorts ..."[8]

"2.3.2 Lack of Novelty due to "Charcoal Production with reduced Emissions" (E2)

E2 describes a "Charcoal production with reduced emissions" (title). This article therefore describes processes for the production of biochar (feature M1).

Under "1. Background" third paragraph, last sentence it states..."[9]

---

[6] European Patent Office, https://register.epo.org/application?documentId=E6JTN9735375DSU&number=EP17769097&lng=de&npl=false, accessed on 09.03.2025.

[7] European Patent Office, https://register.epo.org/application?documentId=E6JTN9735375DSU&number=EP17769097&lng=de&npl=false, accessed on 09.03.2025.

[8] European Patent Office, https://register.epo.org/application?documentId=E6JTN9735375DSU&number=EP17769097&lng=de&npl=false, accessed on 09.03.2025.

[9] European Patent Office, https://register.epo.org/application?documentId=E6JTN9735375DSU&number=EP17769097&lng=de&npl=false, accessed on 09.03.2025.

The disclosure passages of the individual features of the main claim in the documents prejudicial to novelty must be specified.

"2.4 Lack of Inventive Step
2.4.1 Lack of Inventive Step due to D1 and Common General Knowledge
D1 shows in Figure 4 a "Tunnel pyrolysis chamber" into which "wood products" were introduced (features M1 and M2). Furthermore, Figure 4 shows a combustion chamber....
2.4.2 Lack of Inventive Step due to D2 and Common General Knowledge
D2 discloses a process for the production of biochar, in which a biogenic starting material is pyrolyzed (main claim, features M1 and M2). The resulting pyrolysis gases are used as heating gases (dependent claim 1: "...in the heating chamber generator gases are combusted and the resulting combustion gases are used as heating gases for the retort.", features M3 and M4). Furthermore, dependent claim 8 describes that ...."[10]

The lack of inventive step must be substantiated. It is advisable to support the common general knowledge assumed for a person skilled in the art with relevant technical books.

"2.5 Impermissible Extension of the Main Claim
The feature of ***indirect*** *heating of the retorts* ***through a partition wall,*** *i.e., the feature that "by means of the flue gases, heating of the retorts occurs only* ***indirectly through a partition wall"*** *of the main claim can thus ....*
It is therefore an impermissible extension that the indirect heating of the retorts, as disclosed in the originally filed documents, is achieved by the existence of the partition walls."[11]

The alleged inadmissible extension must be justified in detail.

"2.6 Lack of Feasibility of the Main Claim
Claim 1 contains the feature that "the heating of the starting materials located in the retorts by means of the flue gases is effected by heating the retorts only ***indirectly*** *through a partition wall."*
The question arises as to whether the term "indirectly" by ...
The main claim is therefore not feasible for the skilled person."[12]

It must be explained why it is not possible for the skilled person to carry out the teaching of the main claim using their expertise and skills.

"2.7 Dependent Claims to the Main Claim
As is evident directly from documents E1 and E2, dependent claims 2 to 8 are either ...

2.8 Auxiliary Claim 9 and Dependent Claims
Auxiliary claim 9 describes the same features ..."[13]

---

[10] European Patent Office, https://register.epo.org/application?documentId=E6JTN9735375DSU&number=EP17769097&lng=de&npl=false, accessed on 09.03.2025.

[11] European Patent Office, https://register.epo.org/application?documentId=E6JTN9735375DSU&number=EP17769097&lng=de&npl=false, accessed on 09.03.2025.

[12] European Patent Office, https://register.epo.org/application?documentId=E6JTN9735375DSU&number=EP17769097&lng=de&npl=false, retrieved on 09.03.2025.

[13] European Patent Office, https://register.epo.org/application?documentId=E6JTN9735375DSU&number=EP17769097&lng=de&npl=false, retrieved on 10.03.2025.

It must be described where the features of the dependent claims, and in particular the features of the independent claims, can be found in the prior art.

"3. Summary
In this situation, the request for revocation of the contested patent in its entirety is justified. A decision in accordance with the request is sought."[14]

12.4 Example of a Response to an Opposition Brief

The second example explains a response by the patent proprietor to an opposition brief filed by the opponent. An opposition was filed by the opponent OVALO GmbH against European Patent EP 1 972 832 (08 102 558.7) owned by ZF-Lenksysteme GmbH. ZF-Lenksysteme GmbH responded with the following brief.

The contested patent concerns a strain wave gear, also known as a harmonic drive or sliding wedge gear, which is a gear unit with an elastic transmission element and very high transmission ratio and stiffness.

"The patent proprietor comments as follows on the above-mentioned opposition to the above-named patent (contested patent):

1. Requests
It is requested that the opposition be rejected and the contested patent be maintained in its entirety, in view of the reasoning set out below.

Should the Opposition Division come to a view contrary to the rejection of the opposition and maintenance of the contested patent in its entirety, an opportunity is requested to submit at least one further written statement on behalf of the patent proprietor.

Alternatively, and in the event that the Opposition Division comes to a view contrary to the maintenance of the contested patent in whole or in part, an oral hearing pursuant to Article 116 EPC is requested."[15]

An oral hearing should always be requested in order to preserve the opportunity to present one's arguments orally.

"2. Reasoning
2.1 Subject Matter of Claim 1

The preamble of claim 1 describes a strain wave gear for an actuator of a steering system, in particular for a power steering system of a motor vehicle, with an eccentric drive core that can be driven by a servomotor, which elastically deforms a radially flexible rolling sleeve in the radial direction, wherein a toothing on an outer circumferential surface of the radially flexible rolling

[14] European Patent Office, https://register.epo.org/application?documentId=E6JTN9735375DSU&number=EP17769097&lng=de&npl=false, retrieved on 09.03.2025.

[15] European Patent Office, https://register.epo.org/application?documentId=ETTAHBJ43027FI4&number=EP08102558&lng=de&npl=false, retrieved on 08.03.2025.

sleeve partially engages in continuous alternation with a rigid internal toothing of a support ring, and wherein the internal toothing of the support ring is crowned in the direction of the toothing on the outer circumferential surface of the radially flexible rolling sleeve.

The invention consists in the fact that the tooth crest height of the internal toothing increases and decreases by about 0.05 mm to 0.3 mm over the tooth width.

The object leading to claim 1 is described in paragraph [0006], namely to avoid noise generation that can result from coaxial tolerances of the support ring with respect to the radially flexible rolling sleeve of strain wave gears. The aim is therefore to achieve compensation of these manufacturing-related coaxial tolerances in their causation of noise generation. According to the invention, this can be achieved by a tooth crest height profile with a rise and fall of at least 0.05 mm and at most 0.3 mm.

The dimensional specifications of the characterizing feature of claim 1 are a consequence of the typical coaxial tolerances. These typical tolerances are generally in the tenth or hundredth of a millimeter range, regardless of whether small or large components are manufactured. Consequently, the same dimensional specifications of the characterizing feature apply to a steering system of a microcar as well as to the steering system of a 60 t articulated truck.

The characterizing feature with its range of 0.05 mm to 0.3 mm does not itself constitute manufacturing tolerances. The manufacturing tolerance for producing the tooth crest height profile may be greater or smaller. According to the technical teaching of claim 1, it is solely relevant that, in the end, the rise and fall of the tooth crest height results in a height difference of at least 0.05 mm and at most 0.3 mm. In this way, the coaxial tolerances of the support ring relative to the radially flexible rolling sleeve can be eliminated or at least limited in their effect of causing noise generation."[16]

The subject matter of the independent claims should be interpreted in light of the description of the patent in order to set appropriate accents for distinguishing from the prior art cited by the opponent right from the outset.

"*<u>2.2 Prior Art</u>*

In the present opposition, the following documents are cited (the designations are taken from the statement of grounds for opposition):

E1 Dubbel, Handbook of Mechanical Engineering, 19th edition
E2 Heinz Linke, Spur Gear Toothing
E4 DE 11 2004 002 907 T5
E5 JPA 1993-209655
E6 EP 0 745 786 B1
E7 JPA 1999-264448
E8 JPA 2006-15865
E9 WO 2004/000629 A2
E10 WO 2005/110833 A1
E11 Measurement protocol regarding roughness

The following documents are cited in the European search report prepared for the patent application underlying the contested patent:

[16] European Patent Office, https://register.epo.org/application?documentId=ETTAHBJ43027FI4&number=EP08102558&lng=de&npl=false, accessed on 09.03.2025.

D1 SU 1 106 938 A1
D2 US 2002/184968 A1
D3 English translation of JP 58 042452 U
D4 GB 1 128 205 A
E3 WO 2005/054035"[17]

It is advisable to provide a list of the prior art documents with their respective abbreviations (D1, D2, … or E1, E2, …). This helps to maintain an overview.

In the response to a notice of opposition, a feature breakdown of the set of claims is typically not included, since the opponent has already performed a feature analysis, which is generally adopted in order to avoid confusion due to differing feature designations.

"The novelty and inventive step of claim 1 with respect to documents D1 to D4 and document E3 were acknowledged in the European examination procedure.
The novelty of claim 1 with respect to each of the documents E1 to E11 and D1 to D4 is undisputed.
The following summarizes the technically relevant content of the prior art with respect to the strain wave gear according to claim 1 of the contested patent."[18]

It is advisable to interpret the cited documents representing the state of the art, for example with regard to which problem is solved by the respective technical teaching, in order to thereby demonstrate a distinction between the claims of the contested patent and the prior art.

"2.2.1 English Translation of JP 58 042452 U (D3)
D3 discloses a strain wave gear with an internal toothing 12, on which a mechanism A is arranged (Figs. 1, 3, and 5). Mechanism A is intended to reduce the diameter of the internal toothing 12 (page 6, paragraph 1) and thereby achieve a deformation of the internal toothing.

D3 describes that backlash between gears can result in rattling. This rattling is to be reduced by the deformation (page 3, paragraph 3). The objective of D3 is to reduce the backlash in a cost-effective manner (page 3, last paragraph to page 4, paragraph 1). The solution to this is mechanism A.

D3 teaches to tighten screw 19 of mechanism A in order to achieve increasing deformation of the internal toothing. As a result, a central area of a tooth is pressed out, while the edge areas of the tooth remain recessed, which leads to a crowning of the individual tooth.

D3 therefore addresses the problem of noise generation due to gear backlash. The solution is the elimination of the backlash (page 5, paragraph 2).

In contrast, the technical teaching of the contested patent deals with the problem of coaxial tolerances of the support ring relative to the radially flexible rolling bush of strain wave gears [0006]. The appropriate solution for this, namely the rise and fall of the tooth crest height of at least 0.05 mm and at most 0.3 mm, cannot be derived from D3 either.

[17] European Patent Office, https://register.epo.org/application?documentId=ETTAHBJ43027FI4&number=EP08102558&lng=de&npl=false, accessed on 09.03.2025.

[18] European Patent Office, https://register.epo.org/application?documentId=ETTAHBJ43027FI4&number=EP08102558&lng=de&npl=false, accessed on 09.03.2025.

2.2.2 SU 1 106 938 A1 (D1)
D1 describes a strain wave gear with an internal toothing of a support ring, wherein the internal toothing runs monotonically decreasing (Figure 1). There is no rise and subsequent fall across a tooth width shown."[19]

Figure 12.5 shows a strain wave gear of SU 1 106 938 A1.

"2.2.3 US 2002/184968 A1 (D2)
D2 deals with problems that arise in lightweight strain wave gears [0006 to 0009]. Document D2 does not address the problem of reducing noise generation and provides no indications in this regard. Fig. 3 also does not show a continuous rise and subsequent fall of a tooth crest height. Instead, rounded corners and a central straight section of a tooth are depicted."[21]

Figure 12.6 shows a strain wave gear of US20020184968A1.

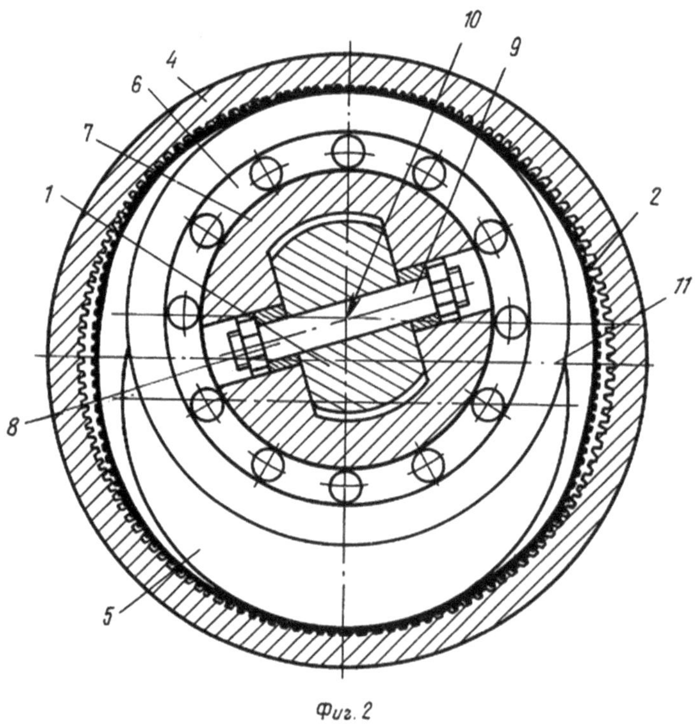

Fig. 12.5 Fig. 2 of SU1106938A1[20]

[19] European Patent Office, https://register.epo.org/application?documentId=ETTAHBJ43027FI4&number=EP08102558&lng=de&npl=false, accessed on 09.03.2025.

[20] DPMA, https://depatisnet.dpma.de/DepatisNet/depatisnet?action=pdf&docid=SU000001106938A1&xxxfull=1, accessed on 10.03.2025.

[21] European Patent Office, https://register.epo.org/application?documentId=ETTAHBJ43027FI4&number=EP08102558&lng=de&npl=false, accessed on 09.03.2025.

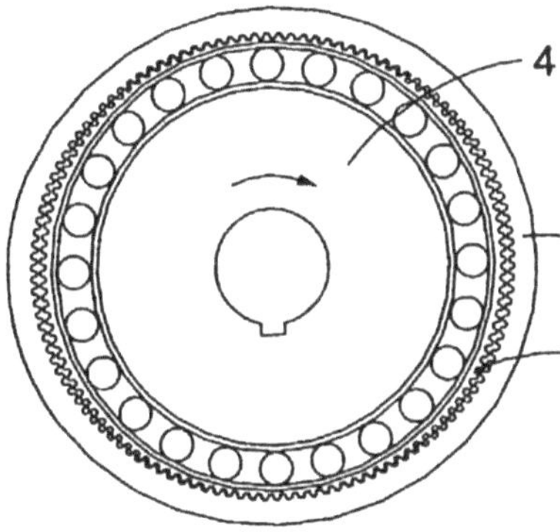

Fig. 12.6 Fig. 2 of US20020184968A1 (DPMA, https://depatisnet.dpma.de/DepatisNet/depatisnet?action=pdf&docid=US020020184968A1&xxxfull=1, accessed on 21.5.2025)

"2.2.4 DE 11 2004 002 907 T5 (E4)
E4 addresses the problem of insufficient toothing depth in a strain wave gear. During high-load torque, the phenomenon of tooth skipping ("ratcheting") can occur (paragraph [0004]). To solve this problem, a strain wave gear is proposed with a circular, rigid internal gear or internally toothed gear, a flexible external gear or externally toothed gear located therein, and a wave generator inserted therein, wherein the flexible external gear has a flexible cylindrical main part and an annular membrane (paragraph [0005]). A tooth profile of the internal gears is described, in which the area of the tooth profile of the rigid internal gear can be greatest where its tooth crest or tooth crest circle touches the maximum point of the movement path of the tooth crest or tooth crest circle of the flexible external gear (paragraph [0006]). E4 does not address noise generation."[22]

Figure 12.7 shows a strain wave gear from DE112004002907T5.

"2.2.5 JPA 1993-209655 (E5)
The arguments of the opponent essentially refer to Figure 5. This figure shows a "cup-shaped" tooth with beveled flanks and a straight central section. The issue of noise generation is not addressed.

2.2.6 EP 0 745 786 B1 (E6)
E6 deals with the problem of tooth collision and proposes a conical design of the individual tooth as a solution to this problem (claim 14). E6 does not address the issue of noise generation."[23]

Figure 12.8 shows a strain wave gear from EP0745786B1 in a perspective view.

Figure 12.9 shows the strain wave gear of EP0745786B1 in a front view.

In the further reasoning as to why the opposition should be dismissed, reference can be made to the brief summaries of the prior art documents. In particular, the could-would test can be used to argue for the inventive step of the independent claims of the contested patent.

[22] European Patent Office, https://register.epo.org/application?documentId=ETTAHBJ43027FI4&number=EP08102558&lng=de&npl=false, accessed on 09.03.2025.

[23] European Patent Office, https://register.epo.org/application?documentId=ETTAHBJ43027FI4&number=EP08102558&lng=de&npl=false, accessed on 09.03.2025.

FIG. 1

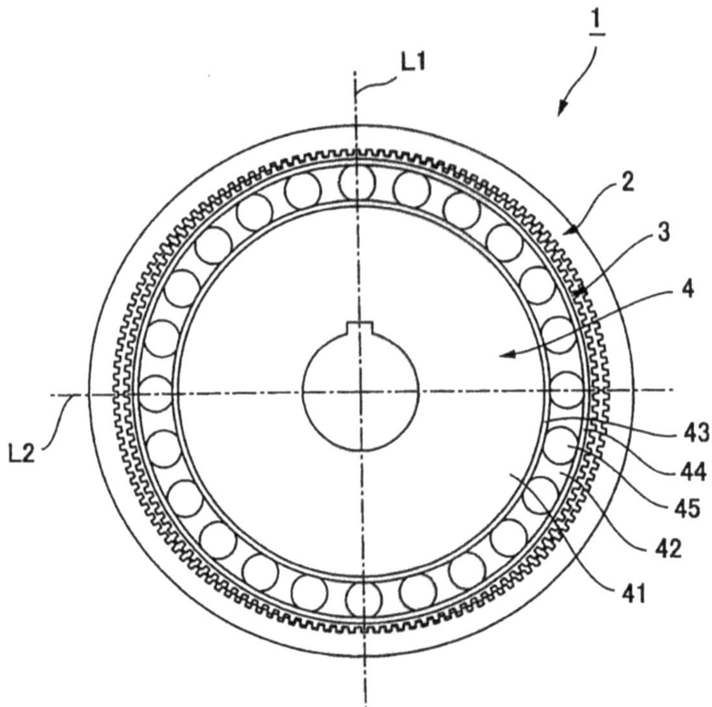

Fig. 12.7 Fig. 1 of DE112004002907T5 (DPMA, https://depatisnet.dpma.de/DepatisNet/depatisnet?action=pdf&docid=DE112004002907T5&xxxfull=1, accessed 21.5.2025)

"2.3 Inventive Step

None of the documents of the prior art recognizes the problem arising from the coaxial tolerances between the support ring and the radially flexible rolling bush of strain wave gears. Consequently, the resulting undesirable noise generation is not described anywhere.

D3 describes the problem of noise generation. However, D3 addresses gear backlash (page 3, paragraph 3 of the translation of the Japanese publication) as the source of undesirable noise. The solution offered by D3 is to eliminate the gear backlash (page 5, paragraph 2 of the translation of the Japanese publication).

The object of the contested patent, which leads to the subject matter of claim 1, is therefore not addressed in any of the documents of the prior art, namely to avoid or at least minimize undesirable noise generation caused by coaxial tolerances between the support ring and the radially flexible rolling bush of strain wave gears. Nor can a connection be constructed between gear backlash and coaxial tolerances, since gear backlash may be intentional, for example to facilitate assembly or to reliably prevent jamming of intermeshing gears. Coaxial tolerances, on the other hand, are not intentional but are due to manufacturing.

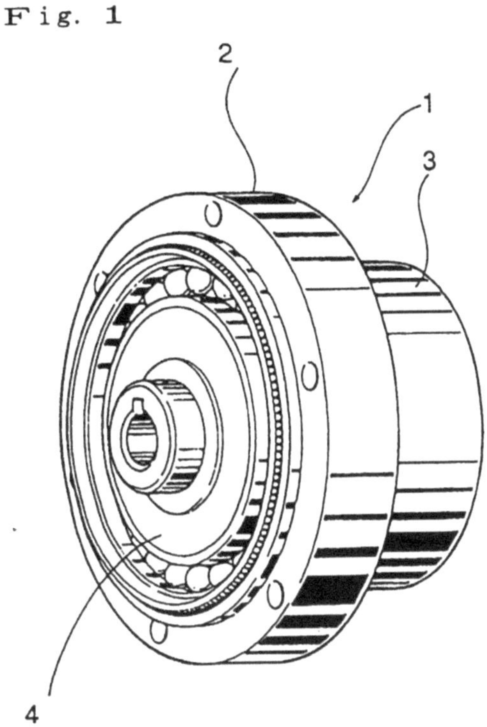

Fig. 12.8 Fig. 1 of EP0745786B1 (DPMA, https://depatisnet.dpma.de/DepatisNet/depatisnet?action=pdf&docid=EP000000745786B1&xxxfull=1, accessed on 21.5.2025)

The inventive solution for avoiding noise generation due to coaxial tolerances also cannot be derived from any of the documents of the prior art. A tooth crest height profile that rises and falls again within the range of 0.05 mm to 0.3 mm is neither described nor suggested anywhere.

A key insight of the technical teaching of the contested patent is that by varying the tooth crest height profile of the internal gearing of the support ring within the range of typical manufacturing tolerances, the influence of the coaxial tolerances of the support ring relative to the radially flexible rolling bush of strain wave gears on noise generation can be eliminated.

Since neither the object nor its solution can be derived from the documents of the prior art (not even in their entirety), an inventive step must be assumed.

2.4 Unallowable Extension

The system described in claim 1 is a strain wave gear. The servomotor is not part of the strain wave gear system. Therefore, a formulation of an eccentric drive core that is "driven by a servomotor" cannot be used.

On page 1, paragraph 2 of the original application documents, a strain wave gear is described that has a drive core. A servomotor sets this drive core in rotation and thus drives it. The drive core of the strain wave gear can therefore be driven by the servomotor. An unallowable extension of claim 1 therefore cannot be seen.

Fig. 8

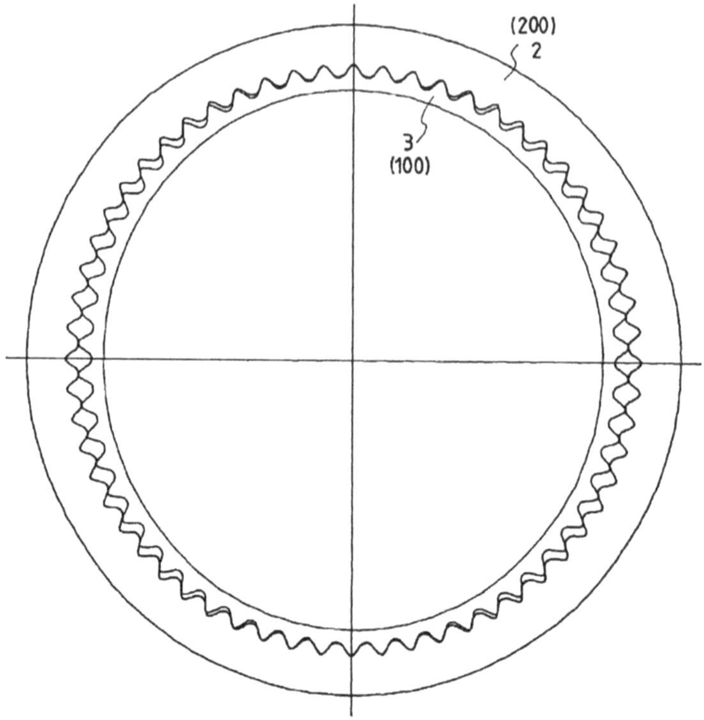

Fig. 12.9 Fig. 8 of EP0745786B1 (DPMA, https://depatisnet.dpma.de/DepatisNet/depatisnet?action=pdf&docid=EP000000745786B1&xxxfull=1, accessed on 21.5.2025)

2.5 Insufficient Disclosure
The subject matter of claim 1 is clearly and fully disclosed, which is not disputed by the opponent. The opponent sees in the combination of the features of claim 1 with the features of claim 2 a contradiction.

The rise and fall of the tooth crest height according to the characterizing feature of claim 1 represents a profile shift. Claim 2 claims a variant of a profile shift, namely a positive profile shift. The features of claims 1 and 2 are therefore combinable."[24]

In the response to the notice of opposition, every point of attack raised by the opponent must be addressed. As a rule, the opponent will allege lack of novelty, lack of inventive step, lack of feasibility (insufficient disclosure), and unallowable extension.

[24] European Patent Office, https://register.epo.org/application?documentId=ETTAHBJ43027FI4&number=EP08102558&lng=de&npl=false, retrieved on 09.03.2025.

"3. Summary

For the reasons stated above, the patent proprietor is of the opinion that the subject matter of claim 1 is legally valid.

Accordingly, the request to dismiss the opposition and to maintain the contested patent in its entirety is justified."[25]

[25] European Patent Office, https://register.epo.org/application?documentId=ETTAHBJ43027FI4&number=EP08102558&lng=de&npl=false, retrieved on 09.03.2025.

European Opposition Appeal 13

Contents

An opposition appeal is the legal remedy against the decision of an opposition division. The appeal is heard before a Board of Appeal of the European Patent Office, which constitutes the final instance in opposition proceedings.

13.1 Timeline of an Opposition Appeal

Figure 13.1 shows the chronological sequence of an opposition appeal, whereby in the written procedure the parties are given at least two opportunities to respond or to comment on the arguments of the opposing party.

T. H. Meitinger, *Grant of a Patent, Opposition, Appeal, and Invalidity*,
https://doi.org/10.1007/978-3-662-72556-6_13

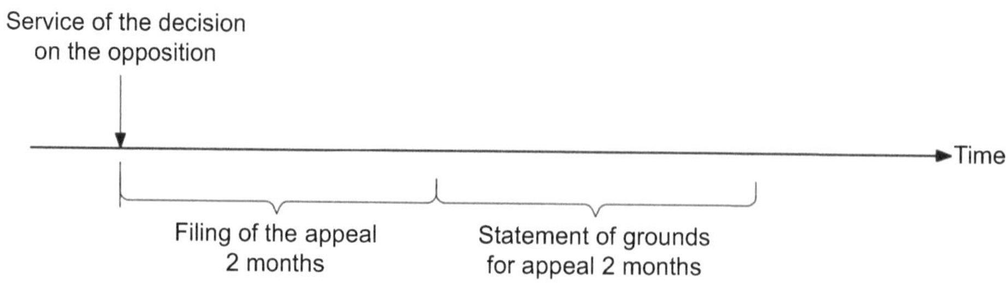

Fig. 13.1 Timeline of an opposition appeal procedure

13.2 Appealable Decision

The decisions of the Receiving Section of the European Patent Office, the examining divisions, and the Legal Division may be challenged by appeal. In particular, the decisions of the opposition divisions are subject to appeal.

13.3 Boards of Appeal

The Boards of Appeal constitute the first and, as a rule, the final judicial instance for the review of decisions of the European Patent Office.[1]

13.4 Time Limit and Form

An appeal may be filed in writing with the European Patent Office within two months of notification of the decision on the opposition. An appeal is deemed to have been filed only if the appeal fee is paid within the same two-month period.[2] The grounds for the appeal must be submitted within a further two months.[3]

[1] European Patent Office, https://www.epo.org/de/case-law-appeals/procedure, accessed on 27.02.2025.

[2] Article 108, sentences 1 and 2 of the European Patent Convention.

[3] Article 108, sentence 3 of the European Patent Convention.

13.5 Review by the Enlarged Board of Appeal

If a party to an opposition appeal is adversely affected, it may, in exceptional cases, seek a review of the decision of the appeal proceedings. A prerequisite for this is, in particular, that the Board of Appeal was composed incorrectly[4], the principle of the right to be heard was violated[5], or a serious procedural defect occurred during the appeal proceedings[6].

In proceedings before the Enlarged Board of Appeal, the parties are given sufficient opportunity in the written part to present their legal views.[7] For this purpose, before the oral proceedings, the Board may, in a non-binding communication, draw the parties' attention to the points to be clarified.[8]

As a rule, an oral hearing is held in which the parties may present their arguments. If the case is ready for decision, the chairman of the Board decides immediately after the oral proceedings and announces the decision (decision from the chair).[9]

[4] Article 112a(2)(b) European Patent Convention; Article 3(1) of the Rules of Procedure of the Boards of Appeal.

[5] Article 112a(2)(c) in conjunction with Article 113(1) European Patent Convention.

[6] Article 112a(2)(d) European Patent Convention.

[7] Article 14(1) Rules of Procedure of the Enlarged Board of Appeal.

[8] Article 14(2), first sentence, in conjunction with Article 13 Rules of Procedure of the Enlarged Board of Appeal.

[9] Article 14(6) Rules of Procedure of the Enlarged Board of Appeal.

Patent Infringement Proceedings 14

Contents

With an infringement action, the patent holder takes legal action against infringements of their patent. A typical reaction of the alleged patent infringer is to file a nullity action against the relevant patent before the Federal Patent Court.

In response to a patent infringement, the patent holder can issue a request for justification, a warning letter, seek a preliminary injunction, or initiate legal proceedings. A request for justification gives the alleged patent infringer the opportunity to present arguments to the patent holder that speak against a patent infringement. A warning letter represents the final attempt to settle a patent dispute out of court. A warning letter always contains the threat of legal action and the demand for a cease-and-desist declaration subject to penalty. A preliminary injunction can be sought in urgent cases, and filing an action for injunctive relief against the patent infringer constitutes patent enforcement, which is considered the standard procedure by the legislator.

T. H. Meitinger, *Grant of a Patent, Opposition, Appeal, and Invalidity*,
https://doi.org/10.1007/978-3-662-72556-6_14

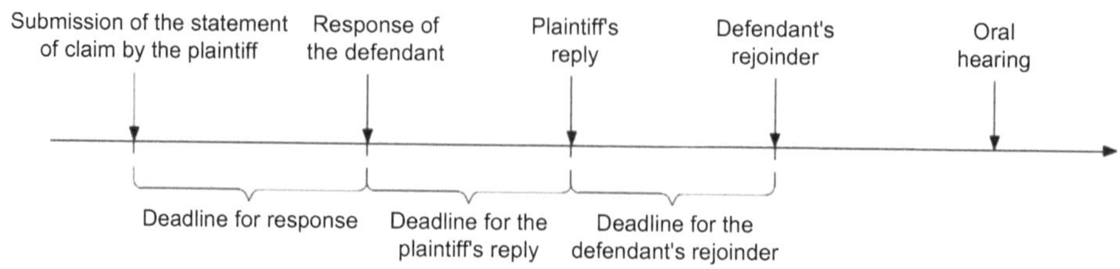

Fig. 14.1 Timeline of a patent infringement proceeding

14.1 Timeline of a Patent Infringement Proceeding

The infringement proceedings begin with the filing of a statement of claim by the patent holder at a competent regional court. The statement of claim is served on the defendant by the court, which grants the defendant a deadline to respond. The patent holder can reply to this response from the defendant with a further statement, the rejoinder. The defendant is then granted another opportunity to respond, the surrejoinder, so that the parties' arguments are sufficiently exchanged. This prepares the subsequent oral hearing, and the panel in charge is able to familiarize itself with the parties' arguments. After the conclusion of the oral hearing, the court renders its judgment. Figure 14.1 shows the timeline of the patent infringement proceedings before a civil court.

14.2 Claims

The patent holder or an exclusive licensee is entitled to a number of claims that can be enforced in infringement proceedings.

Claim for Injunctive Relief

A claim for injunctive relief exists in the case of a risk of recurrence if the patented invention has already been used by an unauthorized party and a patent infringement has therefore occurred.[1] A claim for injunctive relief also arises in the case of an imminent first infringement, if a first patent infringement is likely to occur soon.[2]

[1] Section 139(1) sentence 1 Patent Act.

[2] Section 139(1) sentence 2 Patent Act.

Damages

The patent holder or exclusive licensee may claim damages in the event of intent or negligence, i.e., fault, on the part of the patent infringer.[3] Negligence is already present if no search for third-party patents was conducted prior to using an invention. Therefore, a claim for damages is generally justified, as it is evident that a search was neglected. However, if a search was conducted and the patent was still infringed, intent may even be presumed.

Damages can be calculated according to three different methods, with the choice of method being at the discretion of the patent holder. In particular, damages can be calculated based on the license analogy. This method is the one most commonly applied. In addition, the patent holder can be compensated for lost profits, or the patent holder may receive the infringer's profits.

Destruction

The patent holder can demand that infringing products be destroyed.[4] If the patent claims a process, the claim for destruction relates to products that are directly obtained from the use of the patented process.[5] If special devices and materials used to manufacture infringing products are owned by the patent infringer, the claim for destruction can also extend to these devices and materials.[6] The principle of proportionality must be observed in the claim for destruction, and the legitimate interests of third parties must be taken into account.

Information

The patent holder is entitled to a claim for information, so that they can gain an understanding of the extent of the patent infringement.[7] This also enables them to quantify the damages claimed. The claim for information relates to the suppliers and the distribution channel of the infringing products.

Recall

The patent holder is entitled to a claim for recall, so that infringing products can be permanently removed from the distribution chain.[8]

---

[3] Section 139(2) sentence 1 Patent Act.

[4] Section 140a(1) sentence 1 Patent Act.

[5] Section 140a(1) sentence 2 Patent Act.

[6] Section 140a(2) Patent Act.

[7] Section 140b(1) Patent Act.

[8] Section 140a(3) sentence 1 Patent Act.

14.3 Active Legitimation

The claims can be enforced by the patent holder and the exclusive licensee. A simple licensee is not entitled to do so on their own. However, the simple licensee can be actively legitimated by means of voluntary procedural standing.[9]

14.4 Limits of Enforcement of Claims

Enforcement of claims arising from a patent is excluded in the case of private use.[10] In addition, the individual preparation of a medicinal product by a pharmacist cannot be prohibited.[11] The experimental investigation of a patented invention is permitted if the aim is to gain knowledge about the subject matter of the invention.[12] However, this does not apply if the experiments do not relate to the patented teaching itself, but the patented technical teaching merely serves as a tool to gain other types of knowledge.[13]

Ships, aircraft, or automobiles that contain infringing products but have only temporarily or accidentally entered German territory cannot be considered patent infringements.[14]

The enforcement of a patent can be prevented by a prior user right, provided that the beneficiary of the prior user right was already using the invention at the time of the patent application or had at least created the necessary conditions for its use. If the patent claims a priority, the prior use must have occurred correspondingly earlier.[15]

14.5 Patent Litigation Chambers

For each federal state in Germany, so-called "patent litigation chambers" have been designated, which have exclusive jurisdiction over patent disputes for the respective state. Patent litigation chambers are established in particular in Düsseldorf, Mannheim, and Munich.[16]

---

[9] Mes, 5th ed. 2020, PatG Section 139 para. 44–51; Osterrieth, Part 6. Patent Infringement Osterrieth, Patent Law, 6th edition 2021, para. 927.

[10] Section 11 no. 1 Patent Act.

[11] Section 11 no. 3 Patent Act.

[12] Section 11 no. 2 Patent Act.

[13] Federal Court of Justice 11.07.1995—X ZR 99/92, GRUR 1996, 109—Clinical Trials I; Federal Court of Justice 17.04.1997—X ZR 68/94, NJW 1997, 3092—Clinical Trials II.

[14] Section 11 no. 4 and 5 Patent Act.

[15] Section 12(1) sentence 1 Patent Act.

[16] Section 143 paragraph 2 sentence 1 Patent Act.

14.6 Mandatory Legal Representation

Before the Regional Court and the Higher Regional Court, legal representation by an attorney is mandatory.[17] This means that both the patent holder and the alleged patent infringer must be represented by a lawyer. A patent attorney is not authorized to represent clients before an ordinary court.

14.7 Costs

Before the ordinary courts, the principle of losing party pays applies, so that the losing party is required to bear all costs, including court costs and the costs of the opposing party.[18]

[17] Section 78 paragraph 1 sentence 1 Code of Civil Procedure.
[18] Section 91 paragraph 1 sentence 1 Code of Civil Procedure.

Unitary Patent 15

Contents

Since September 1, 2024, it has been possible to obtain a unitary patent for currently 18 member states following a grant procedure before the European Patent Office. Alternatively, as before, national European patents can be pursued.

Currently, a joint unitary patent can be sought for the following states: Belgium, Bulgaria, Denmark, Germany, Estonia, Finland, France, Italy, Latvia, Lithuania, Luxembourg, Malta, the Netherlands, Austria, Portugal, Romania, Slovenia, Sweden.[1]

The major advantage of the unitary patent is that validation, monitoring, and payment of annual fees for individual countries are no longer required. The unitary patent therefore reduces administrative complexity.[2]

[1] EPO, https://www.epo.org/de/applying/european/unitary/unitary-patent, accessed on 25.02.2025.
[2] EPO, https://www.epo.org/de/applying/european/unitary/unitary-patent, accessed on 25.02.2025.

T. H. Meitinger, *Grant of a Patent, Opposition, Appeal, and Invalidity*,
https://doi.org/10.1007/978-3-662-72556-6_15

15.1 Procedure for the Unitary Patent

A unitary patent can be obtained by filing a request for unitary effect.[3] The request must be filed within one month after publication of the mention of the grant of the European patent in the European Patent Bulletin. The request must be submitted to the European Patent Office.[4]

The request for unitary effect must include the applicant who is the proprietor of the European patent, the number of the European patent, the representative of the applicant, and a translation of the European patent into English if the language of the proceedings is German or French, or into German or French if the language of the proceedings is English.[5]

The European Patent Office examines the request and grants the unitary patent if the formal requirements are met.[6] If the request was filed in due time but not all formal requirements were met, the applicant or their representative may remedy the formal deficiencies within a period of one month.[7]

15.2 Annual Fees

The annual fees and the surcharge fees for late payment must be paid to the European Patent Office.[8]

The annual fees for a unitary patent are as follows (Fig. 15.1):[9]

2nd year	35 €	12th year	1,775 €
3rd year	105 €	13th year	2,105 €
4th year	145 €	14th year	2,455 €
5th year	315 €	15th year	2,830 €
6th year	475 €	16th year	3,240 €
7th year	630 €	17th year	3,640 €
8th year	815 €	18th year	4,055 €
9th year	990 €	19th year	4,455 €
10th year	1,175 €	20th year	4,855 €
11th year	1,460 €		

Fig. 15.1 Annual fees for a unitary patent

[3] Rule 5(1) Rules relating to Unitary Patent Protection (UPR).
[4] Rule 6(1) Rules relating to Unitary Patent Protection (UPR).
[5] Rule 6(2) Rules relating to Unitary Patent Protection (UPR).
[6] Rule 7(1) Rules relating to Unitary Patent Protection (UPR).
[7] Rule 7(3) Rules relating to Unitary Patent Protection (UPR).
[8] Rule 13(1) sentence 1 Rules relating to Unitary Patent Protection (UPR).
[9] As of February 2025.

15.3 Unified Patent Court

The Unified Patent Court is a supranational patent court responsible for unitary patents and European patents. The Unified Patent Court is particularly responsible for infringement proceedings and proceedings for judicial review of the validity of the relevant patent through an action for nullity. The parties benefit from the advantage that they do not have to act before several national courts, but can obtain a decision for currently 18 states with a single court proceeding.[10]

The Unified Patent Court consists of a court of first instance and a court of appeal. The central division of the first instance is located in Paris with a section in Munich. In addition, there are local and regional divisions of the first instance. The second instance of the Unified Patent Court is the court of appeal.[11] The court of appeal is based in Luxembourg.[12]

---

[10] EPO, https://www.epo.org/de/applying/european/unitary/upc, accessed on 26.02.2025.

[11] Unified Patent Court, https://www.unified-patent-court.org/de/gericht/vorstellung, accessed on 26.02.2025.

[12] EPO, https://www.epo.org/de/applying/european/unitary/upc, accessed on 26.02.2025.

Trademarks 16

Contents

A trademark can be challenged by a timely opposition or a cancellation procedure. Through infringement proceedings before a civil court, the rights of the trademark owner can be asserted against the infringer. In particular, the trademark owner has the right to prohibit unauthorized use of their registered trademark in a trademark-related manner.

16.1 Opposition

German Trademark

Within a period of three months after the publication of the registration of a trademark, the owner of an earlier trademark or a business designation may file an opposition against the registration of the trademark.[1]

Grounds for Opposition

An opposition may only be based on earlier rights, in particular trademark rights, designations of origin, geographical indications, or business designations, or may be directed against the registration of an agent's mark.[2]

[1] Section 42 (1) sentence 1 of the Trademark Act.

[2] Section 42 (2) of the Trademark Act.

T. H. Meitinger, *Grant of a Patent, Opposition, Appeal, and Invalidity*,
https://doi.org/10.1007/978-3-662-72556-6_16

Absolute grounds for refusal, such as lack of distinctiveness and infringement of the need to keep free, do not constitute grounds for revocation in opposition proceedings.

Amicable Settlement
The opposition proceedings can be suspended for at least two months to give the parties the opportunity to reach an amicable settlement. This requires an request from all parties involved in the proceedings.[3]

Content of the Opposition
The opposition must specify the challenged trademark, the earlier right, and the opponent.[4] If an earlier right is asserted that is not recorded in the register of a patent office, the type, representation, form, seniority, subject matter, and owner of the sign right must be indicated.[5]

Suspension
The German Patent and Trademark Office may suspend opposition proceedings if it considers this expedient.[6] Suspension is particularly considered if the opposition is likely to succeed and is based on a pending, not yet registered, trademark or if cancellation proceedings are pending against the opposition mark.[7]

Value in Dispute in Opposition Proceedings
The attorney's fee is based on the value in dispute of the proceedings. A newly registered, young trademark may not yet have acquired significant value. The senates of the Federal Patent Court typically assume a value in dispute of €20,000 to €50,000.[8]

Ownership of Rights
An opposition may only be filed by the owner of an earlier right on the basis of that right.[9] Ownership is presumed for the person entered in the register.[10]

Plea of Non-Use
The owner of the later trademark may raise the plea of non-use if the five-year grace period for use of the opposition mark has already expired after registration. In this case,

[3] Section 42 (4) of the Trademark Act.
[4] Section 30 (1) sentence 1 of the Trademark Ordinance.
[5] Section 30 (1) sentence 2 of the Trademark Ordinance.
[6] Section 32 (1) of the Trademark Ordinance.
[7] Section 32 (2) of the Trademark Ordinance.
[8] BPatG 27 W(pat) 75/08; BPatG 26 W (pat) 47/10, BPatG 24 W (pat) 18/10; BPatG 25 W (pat) 29/10; BPatG 28 W (pat) 52/09; BPatG 30 W (pat) 108/05; BPatG 33 W (pat) 84/04.
[9] Section 42 (1) of the Trademark Act.
[10] Section 28 (1) of the Trademark Act.

the owner of the opposition mark must credibly demonstrate genuine use of their mark within a period of five years prior to the publication of the registration of the mark against which the opposition is directed.[11] To credibly demonstrate means that the disputed use is proven with a preponderance of probability. No formal proof is required.[12]

Withdrawal of the Opposition

The withdrawal of the opposition results in the termination of the opposition proceedings. Only the owner of the opposition mark is entitled to withdraw the opposition.[13] A withdrawal in an appeal procedure against the decision in opposition proceedings renders any possible cancellation of the later trademark ineffective.[14]

European Union Trademark

Within a period of three months after publication of the application for a European Union trademark, opposition to the registration of a European Union trademark may be filed with the EUIPO. An opposition can only be based on the claim that the younger mark infringes the right of an earlier right. An infringement exists if there is a likelihood of confusion between the earlier and the younger mark or if there is double identity. In the case of double identity, it is assumed that the goods and services and the trademark representations are identical.

The opposition must be filed in writing and substantiated. An opposition is only deemed to have been validly filed once the opposition fee has been paid.[15]

Opposition Procedure

The EUIPO invites the parties to the opposition, setting a deadline, as often as necessary to comment on official communications or on submissions from the opposing party, as required to clarify the matter.

Plea of Non-Use

The applicant of the later mark may require the owner of the earlier mark to provide proof of use. In this case, the owner must prove use during the five years prior to the filing or priority date of the contested trademark application. This is only required if the European Union trademark has already been registered for five years.[16]

---

[11] Section 43 (1) of the Trademark Act.

[12] Section 43 (1) of the former version of the Trademark Act in conjunction with Section 294 of the Code of Civil Procedure.

[13] Section 42 (1) of the Trademark Act.

[14] Section 82 (1) sentence 1 of the Trademark Act in conjunction with Section 269 (3) sentence 1 of the Code of Civil Procedure.

[15] Article 46 (3) of the European Union Trademark Regulation.

[16] Article 47 (2) sentence 1 of the European Union Trademark Regulation.

If proof of use is required but not provided, the opposition will be rejected.[17]

If the opposition mark is a European Union trademark and use is proven only for some of the goods and services, the European Union trademark is considered registered in the opposition proceedings only for those goods and services.[18]

If the application is challenged with a national trademark, when raising the plea of non-use, use must be proven only for the respective Member State.[19]

Decision

The application for the European Union trademark is refused for those goods and services for which the opposition is justified.[20] In all other respects, the opposition is rejected.[21] The decision on the opposition will be published once it becomes final.[22]

16.2 Cancellation

A cancellation request can initiate proceedings before the German Patent Office or the EUIPO to remove a trademark from the register.

German Trademark

A German trademark can be removed from the register through cancellation proceedings before the German Patent and Trademark Office.

Cancellation due to non-use

A request for cancellation is successful if the trademark in question has not been used within an uninterrupted period of five years from the date on which opposition was no longer possible.[23] Cancellation is not possible if there has been genuine use in Germany for the registered goods and services.[24] The grounds for cancellation may be remedied if use has been resumed.[25] Remediation does not occur if use commenced after knowledge of the cancellation request.[26]

[17] Article 47 (2) sentence 2 of the European Union Trademark Regulation.

[18] Article 47 (2) sentence 3 of the European Union Trademark Regulation.

[19] Article 47 (3) of the European Union Trademark Regulation.

[20] Article 47(5), first sentence, European Union Trademark Regulation.

[21] Article 47(5), second sentence, European Union Trademark Regulation.

[22] Article 47(6) European Union Trademark Regulation.

[23] Section 49(1), first sentence, German Trademark Act.

[24] Section 26(1) German Trademark Act.

[25] Section 49(1), second sentence, German Trademark Act.

[26] Section 49(1), third sentence, German Trademark Act.

Cancellation is also possible if the trademark has become the common designation for the goods and services and the trademark owner has not taken action against this.[27] A request for cancellation will also be successful if there is a risk of deception associated with the trademark.[28]

The grounds for cancellation must be examined for all goods and services. If the grounds for cancellation apply only to certain goods and services, the trademark is to be cancelled only for those goods and services.[29]

Cancellation due to absolute grounds for refusal

The registration of a trademark can be challenged in particular due to lack of distinctiveness[30] or infringement of the need to keep signs free for general use.[31, 32]

Cancellation due to earlier rights

Cancellation due to earlier rights can be achieved by filing a lawsuit before a civil court[33] or by submitting a request to the German Patent and Trademark Office.[34]

European Union Trademark

A European Union trademark can be challenged by a request for cancellation due to revocation or invalidity. An invalidity procedure can be based on absolute or relative grounds for invalidity.

Cancellation due to revocation

A European Union trademark can be declared revoked through official proceedings before the EUIPO. In addition, the trademark can be declared revoked as a result of a counterclaim before the infringement court.[35] The prerequisite for this is that the European Union trademark has not been put to genuine use within a period of five years, and that no genuine use occurred before the filing of the cancellation action.[36] Use within a period of three months before the filing of the cancellation action is disregarded if the use commenced after knowledge of the cancellation action or a threatened cancellation action.[37]

---

[27] Section 49(2) No. 1 German Trademark Act.
[28] Section 49(2) No. 2 German Trademark Act.
[29] Section 49(3) German Trademark Act.
[30] Section 8(2) No. 1 German Trademark Act.
[31] Section 8(2) No. 2 German Trademark Act.
[32] Section 50(1) German Trademark Act.
[33] Section 55(1), first sentence, German Trademark Act.
[34] Section 51(1), first sentence, German Trademark Act.
[35] Article 58(1) European Union Trademark Regulation.
[36] Article 58(1)(a), first and second clauses, European Union Trademark Regulation.
[37] Article 58(1)(a), third clause, European Union Trademark Regulation.

Another ground for revocation exists if the proprietor, although able to do so, did not prevent the European Union trademark from becoming the common designation for a good or service for which it is registered.[38]

Revocation also threatens if the European Union trademark, through its use, has become misleading.[39]

A European Union trademark can only be revoked for those goods and services for which a ground for revocation has arisen.[40]

Cancellation due to invalidity
A European Union trademark can be declared invalid on absolute and relative grounds for invalidity. The difference is that absolute grounds for invalidity can be invoked by anyone, whereas relative grounds for invalidity can only be invoked by the proprietor of the earlier trademark.

16.3 Infringement Proceedings

Infringement proceedings before a civil court can be used by a trademark owner to assert claims against an infringer.

German Trademark
An infringement can be asserted if the alleged infringer uses a confusingly similar designation as a trademark. Trademark infringement therefore only exists if the challenged designation is used as an indication of origin. The same applies in cases of double identity.[41]

European Union Trademark
Infringement of a European Union trademark is heard before the relevant civil court in the respective Member State of the European Union.

[38] Article 58(1)(b) European Union Trademark Regulation.
[39] Article 58(1)(c) European Union Trademark Regulation.
[40] Article 58(2) European Union Trademark Regulation.
[41] Section 14(2) German Trademark Act.

17 Design Rights

Contents

A design right can be challenged through a nullity procedure. Alternatively, a declaration of invalidity can be achieved by way of a counterclaim in infringement proceedings.[1]

17.1 Nullity Proceedings

The nullity procedure is an official procedure conducted against a German design before the German Patent and Trademark Office and against a Community design before the EUIPO.

German Design

A German design can be cancelled through a nullity procedure before the German Patent and Trademark Office.

[1] Section 33(3) Design Act or Article 24(1) Community Design Regulation (CDR).

T. H. Meitinger, *Grant of a Patent, Opposition, Appeal, and Invalidity*,
https://doi.org/10.1007/978-3-662-72556-6_17

Grounds for Nullity

A design can be declared invalid on the basis of absolute and relative grounds for nullity. Absolute grounds for nullity can be asserted by anyone, whereas relative grounds for nullity can only be asserted by the party whose rights have been infringed.

Absolute grounds for nullity are lack of novelty and lack of individual character.[2] Further absolute grounds for cancellation exist if the design is of a purely technical nature or if the design has a particular configuration solely to allow it to be assembled in a complementary manner with other elements. In addition, violation of public order and morality can be asserted as absolute grounds for nullity.[3]

Relative grounds for nullity arise from the infringement of an earlier right, whereby only the holder of the earlier right is entitled to take action. An earlier right does not have to be a design right. In addition to the infringement of a design right, infringement of a copyright or a trademark right can also constitute a relative ground for nullity.

Course of the Nullity Procedure

The nullity procedure is initiated upon application, which must be accompanied by the asserted grounds for nullity, the facts, and the evidence.[4]

The German Patent and Trademark Office forwards the application to the holder of the challenged design and gives them the opportunity to object to the application for cancellation within a period of one month. If no timely objection is filed, the design right is deleted from the register (default).[5]

If the holder of the design right objects, contentious proceedings before the German Patent and Trade Mark Office commence. An oral hearing may be scheduled if requested by one of the parties or if the German Patent and Trade Mark Office considers it expedient.[6]

As a rule, the German Patent and Trade Mark Office will decide on the costs, applying the principle that the losing party bears the costs. Accordingly, in contentious nullity proceedings concerning a design right, the losing party is burdened with the costs of the proceedings. The losing party must therefore also bear the costs incurred by the opposing party for the proper conduct of the proceedings.[7] If no decision on costs is made, each party bears its own costs.[8]

---

[2] Section 33(1) No. 2 Design Act.

[3] Section 33(1) No. 3 in conjunction with Section 3(1) Nos. 1 to 3 Design Act.

[4] Section 34a(1) Sentences 1 and 2 Design Act.

[5] Section 34a(2) Sentences 1 and 2 Design Act.

[6] Section 34a(3) Design Act.

[7] Section 91(1) Sentence 1 Code of Civil Procedure.

[8] Section 34a(5) Sentence 4 Design Act.

A third party may join the nullity proceedings if they can assert a legal interest. In particular, a legal interest exists if the third party has been warned or sued for infringement based on the disputed design.[9]

An application for a declaration of invalidity may be inadmissible if the applicant is bound by a non-assertion agreement. However, this only applies if the non-assertion agreement is permissible, i.e., it does not have the effect of appreciably restricting trade or competition between EU Member States.[10]

If the design right has already expired, it can still be challenged by an application for a declaration of invalidity in order to prevent the enforcement of rights for the past.[11]

Community Design

A Community design can be declared invalid by a nullity procedure before the EUIPO or by a counterclaim in design infringement proceedings before a Community design court.[12]

The applicant must substantiate their application and pay the fee for a declaration of invalidity.[13]

The nullity procedure before the EUIPO is conducted exclusively in writing. There is no oral hearing or trial. The EUIPO invites the parties to comment as often as necessary to clarify the matter. A deadline is set for each party to respond.[14] An already expired design right can be challenged by a nullity procedure due to the assertion of rights for the past.[15]

International Registration

An international registration can be declared invalid for Germany by a nullity procedure before the German Patent and Trade Mark Office.[16]

17.2 Infringement Proceedings

A design right is an unexamined intellectual property right, as its legal validity is not examined by the relevant patent office prior to registration. Therefore, before bringing an infringement action, the claimant should conduct their own assessment.

[9] Section 34c(1) Design Act.
[10] Article 101 TFEU.
[11] Section 33(5) Design Act.
[12] Article 24(1) CDR.
[13] Article 52(2) CDR.
[14] Article 53(2) CDR.
[15] Article 24(2) CDR.
[16] Section 70(1) Sentence 1 Design Act.

An assessment of the design right prior to filing an infringement action involves several steps. First, the prior art in terms of existing designs must be determined. The design is then broken down into its features and examined for legal validity in light of the prior art. In a final step, based on the feature breakdown, it can be determined whether an actual design infringement exists. The assessment is made from the perspective of an informed user.

Prior Art in Design

Identifying the prior art in design serves to assess the legal validity and the scope of protection that can be accorded to the design right.

The prior art in design includes all designs that were known before the filing or priority date of a design right and are relevant for assessing the legal validity and scope of protection of the design right.

If no relevant prior art can be identified, the design right is granted a broad scope of protection. An alleged infringer should therefore, in their own interest, conduct a search for the prior art of the allegedly infringed design in order to limit the scope of protection of the allegedly infringed design.

If, in particular, very similar designs can be identified in the prior art, the relevant design right will only have a narrow scope of protection. In this case, an infringing form must embody all essential features of a design right for an infringement to be assumed.

If the design right is closer to a design from the prior art than to the alleged infringing form, a design infringement can be ruled out.

Overall Impression

The feature breakdown is a listing of the features of the design that determine the overall impression. Purely technical features are to be removed from the feature breakdown. The remaining features are to be weighted according to their significance.

Informed User

For the assessment of legal validity and design infringement, reference is made to the informed user. The informed user is characterized by a high level of attention. The informed user is not a design expert, but does possess a basic knowledge of designs. The informed user notices differences that would not be apparent to an average consumer.[17]

Example of a Feature Breakdown

The Federal Court of Justice created a feature breakdown for a design right concerning a seating arrangement for a waiting area, for example at an airport. The design right

[17] ECJ, 20.10.2011, C-281/10 P, GRUR 2012, 506—PepsiCo.

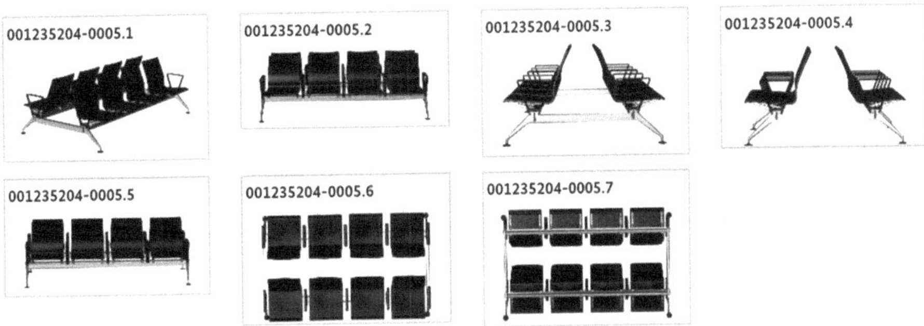

Fig. 17.1 Design right "Waiting Bench" GGM 001235204-0005

in question was the Community Design GGM 001235204-0005. Figure 17.1 shows the seven views of the Community Design.[18]

The feature breakdown of the Federal Court of Justice was as follows:[19]

"Waiting bench

(1)*with a total of 8 continuous, ergonomically shaped, angular seat shells in a dark color,*
(2)*whereby 4 seat shells each, spaced apart, are attached to a horizontal beam by means of two cantilevered struts engaging at the front of the seat shell,*
(3)*the individual seat shells are framed by trapezoidal armrests, the shorter base of which is attached to the beam,*
(4)*the two beams run parallel to each other, so that the two rows of seats are arranged "back-to-back",*
(5)*and are supported only at their two ends by a trapezoidal frame with two feet,*
(6)*whose angled ends have round gliders in the color of the seat shells." (BGH, Decision of 24 January 2019, I ZR 164/17 – Meda Gate, https://juris.bundesgerichtshof.de/cgi-bin/rechtsprechung/document.py?Gericht=bgh&Art=en&sid=30fb1b3707d638be87edcfed2fa5754c&nr=93144&anz=3&pos=2, accessed on 23.5.2025)*

Proceedings for design infringement are held before regional and higher regional courts, with special chambers in each federal state specializing in design infringement cases.[20] In infringement proceedings, the validity of the design underlying the infringement

[18] DPMA, https://register.dpma.de/DPMAregister/gsm/registerhabm?DNR=001235204-0005, accessed on 05.09.2023.
[19] BGH, 24.01.2019, I ZR 164/17—Meda Gate, https://juris.bundesgerichtshof.de/cgi-bin/rechtsprechung/document.py?Gericht=bgh&Art=en&Datum=Aktuell&Sort=12288&nr=93144&pos=27&anz=583, accessed on 05.09.2023.
[20] Section 52 paragraphs 2 and 3 Design Act.

proceedings can be challenged by way of a counterclaim.[21] If such a counterclaim is filed, the court not only examines the infringement but also the legal validity of the design right.

Limits of Enforcement

There are limits to the enforcement of a design right. In particular, prior user rights and the lack of enforceability in the case of a purely technically determined design or a lack of visibility of the design must be taken into account.

A prior user right exists if the design was already used by a third party before its registration with the Patent Office.[22]

In the case of a technically determined configuration of the design, the relevant design right is not legally valid.[23] Technical configurations are intended to be protected by a patent or utility model, not by design law.

Design protection can only be sought for configurations that are visible in normal use.[24] Normal use does not include repair.

[21] Section 52b paragraph 1 sentence 1 Design Act.

[22] BGH, 29.06.2017, I ZR 9/16—Bed Frame, https://juris.bundesgerichtshof.de/cgi-bin/rechtsprechung/document.py?Gericht=bgh&Art=en&az=I%20ZR%209/16&nr=80242, accessed on 06.09.2023.

[23] Section 3 paragraph 1 no. 1 Design Act or Article 8 paragraph 1 CDR.

[24] Section 4 Design Act or Article 4 paragraph 2 lit. a CDR.

MIX
Papier aus verantwortungsvollen Quellen
Paper from responsible sources
FSC® C105338

If you have any concerns about our products,
you can contact us on
ProductSafety@springernature.com

In case Publisher is established outside the EU,
the EU authorized representative is:
Springer Nature Customer Service Center GmbH
Europaplatz 3, 69115 Heidelberg, Germany

Printed by Libri Plureos GmbH
in Hamburg, Germany

Powertrain Development with Artificial Intelligence

Aras Mirfendreski

Powertrain Development with Artificial Intelligence

History, Work Processes, Concepts, Methods and Application Examples

Springer

Aras Mirfendreski
Köln, Germany

ISBN 978-3-662-63862-0 ISBN 978-3-662-63863-7 (eBook)
https://doi.org/10.1007/978-3-662-63863-7

Translation from the German language edition: Künstliche Intelligenz für die Entwicklung von Antrieben by Aras Mirfendreski, © Der/die Herausgeber bzw. der/die Autor(en), exklusiv lizenziert durch Springer-Verlag GmbH, DE, ein Teil von Springer Nature 2022. Published by Springer Berlin Heidelberg. All Rights Reserved.

Responsible Editor: Markus Braun
This Springer imprint is published by the registered company Springer-Verlag GmbH, DE
part of Springer Nature.
The registered company address is: Heidelberger Platz 3, 14197 Berlin, Germany

Contents

List of Figures

List of Tables

1 Introduction

When did we actually start to subject everything around us and not least ourselves to continuous optimization? The idea behind optimization has been around for as long as human existence. To reduce physical labor, to increase the speed of an action in order to shorten its duration or to increase yield and thus the efficiency of an action is a sign of intelligence that is deeply anchored in our thought processes. In the eighteenth century, the agricultural revolution arose out of a necessity to feed the growing population of Western Europe. As a result, innovative processes had to be developed to increase the fertility of the soil and increase the yield of harvest. Evolutionary breeding behaviors helped to raise more powerful farm animals to support agriculture or to breed animals with a higher meat content for consumption. This era went hand in hand with the industrial revolution of the nineteenth and twentieth centuries. The development of power machines and drives changed the attitude toward physical labor. New working communities, as created by industries, as well as new and closer societies emerged, which promoted social networking. This led to the rise of the proletariat for the first time, which in a broader sense was also motivated by the optimization of social structures.

Our natural drive for development has taken us to a highly efficient and highly technological level in which we live today. Once a system has reached higher orders of optimization, it is inevitable to switch to alternative systems in order to conquer new business areas and reposition in the highly dynamic and competitive market. Does this perhaps explain why the combustion engine is experiencing a renaissance?

Drive concepts of the future will have to meet the granular demands of society as well as the strict requirements of politics. Based on demanding expectations, the entire automotive industry is collectively preparing to set the right course at an early stage for a visionary future landscape. Recent publications show that by 2040, a variety of hybrid variants, with a share of $53\% - 65\%$, should cover the new registration market. The concepts can primarily be divided into mild 48V hybrids with different powertrain variants ($P_0 - P_4$ topology), conventional serial or parallel hybrids and plug-in hybrids up to range extenders, where the

A. Mirfendreski, *Powertrain Development with Artificial Intelligence*,
https://doi.org/10.1007/978-3-662-63863-7_1

engine takes over a passive role of the drive. The range 11% – 15% of the market share is to be covered by pure electric vehicles, and the remaining 22% – 29% will be split between pure internal combustion engines, which will be powered to a higher degree by synthetic fuels (e-fuels) and bio-fuels, as well as gas and hydrogen-powered vehicles.

As a result of the targeted and versatile drive variants, the development costs of automobile manufacturers will increase dramatically in the coming years. Combustion engines have already reached an almost maximum potential regarding their efficiency and thus a minimum of CO_2 emissions. Further optimization attempts are now only bringing marginal improvements at the same or even higher investment volume. This, however, does not mean that it can be replaced so easily. Yet, one thing is certain: investment trends for the development of internal combustion engines will no longer be sustained in this form, as they have been in recent decades. An investment basis resulting from more versatile drive concepts will only be profitable in the future if significant leaner development processes and more efficient methods replace the current ones.

For this purpose, the right tools are available at the right time: Artificial intelligence (AI). AI is the science of intelligent agents combined with the power of neural networks that primarily computer scientists have dealt with over the past decades. Their versatile fields of application have settled in the world of IT and have significantly advanced the digitization process since the 1970 s. In the meantime, the concepts of AI have become versatile, reliable and highly efficient. How can the topics of a digital IT sector be transferred to industrial fields?

The term artificial intelligence is used in an inflationary way in the vernacular and meanwhile rather conveys an image or a feeling that a certain process can be carried out cleverly, mechanically and by self-learning. This book is intended as a concept book and serves to consolidate the possible uses of AI for industrial applications. It aims to present AI topics in all its complexity and to provide conceptual and creative ideas on how it can be used in powertrain development. Engineers with a focus on drive development who want to get to know more about the possibilities and levels of AI should benefit from this. This book is also aimed at AI users and IT experts who do not have a classical engineering background and would like to gain a clear impression of how their techniques can be applied to specific powertrain development topics.

Aras Mirfendreski

The Combustion Engine at the Turn of Industrialization

2

2.1 The Prehistory

Industrialization changed the world. It began in the late eighteenth century and brought about a revolutionary change through technology. The abrupt change challenged centuries-old structures at various levels that were unprepared. It seemed to possess extraordinary power and revolutionized agriculture, trade, transport, textile production, housing, politics, cultures and last but not least the entire society.

The need for industrialization was first determined by the agricultural revolution. This took place between 1760 and 1815 and enabled farmers to achieve significant growth in food production through the modernization of technologies in agriculture and animal husbandry. Western Europe and especially Great Britain were traditional trading regions at that time. The combination of ever-increasing population density and poor arable land forced farmers and peasants to become more efficient and to develop new methods to increase their productivity. In arable farming, grazing livestock was used as a fertilizer supplier for pasture land, which increased the fertility of cereal fields. Great successes in the efficient management of fields were achieved by growing new crops that simultaneously served the food cycle between humans and livestock, and by cultivating new and more robust grass varieties that not only grew faster but also extracted fewer nutrients from the soil.

In the field of **agricultural technologies**, the seed drill was invented at the beginning of the eighteenth century, replacing the plowing and sowing by human labor. This machine not only saved a considerable amount of time when sowing fields but also made it easier to harvest with the precision of evenly spaced seeds and hence increased efficiency. Other technical innovations included a horse-drawn hoe, introduced by Jethro Tull around 1708, which was used for weeding, and a lathe by Andrew Meikle around 1760, which separated the grain from its husk.

In the **livestock** sector, Robert Bakewell introduced groundbreaking innovations. For grassland management, he worked close to the river, diverting waterways and canals to

A. Mirfendreski, *Powertrain Development with Artificial Intelligence*,
https://doi.org/10.1007/978-3-662-63863-7_2

irrigate the land. This enabled him to ensure that crops did not fail even in dry periods and to eliminate the dependence on rain. He experimented with different animal fertilizers on trial fields and promoted the fertility of the soil. And in the field of animal breeding, he moved away from uncontrolled breeding of cattle on free pasture. He deliberately brought together healthy and strong sheep, cattle and cows, and, through evolutionary breeding behavior, from generation to generation, was able to raise more breed animals with a higher portion of meat.

The agricultural revolution of the seventeenth century and the subsequent industrial revolution in the late eighteenth and nineteenth centuries were pioneered by Britain, which served as a model for Western Europe, the United States and, in the late nineteenth century, Japan. Britain's rapidly growing population, high supply of labor, involvement in world trade and availability of raw materials were important prerequisites—all of which arrived at the same time, making technological change both necessary and possible.

The driving raw material of industrialization was coal, not only because it has an energy density up to four times higher than wood, but also because it burns much more slowly and thus releases its energy more moderately. Great Britain had large coal deposits. Unlike in many other countries, its coal was not deep, so it was accessible without the need for technologies for deep mining. This and the lack of alternatives to coal, as the country had only a small amount of forested land to supply wood for industries with high energy needs, catapulted Britain forward and gave it a great developmental advantage over other countries.

Toward the end of the eighteenth century, Scottish engineer James Watt developed an efficient version of the steam engine. In the subsequent decades, it found its way into mining, metal and textile industries as a reliable machine. This technology enabled mechanized mining of coal lay in deeper layers of the earth such as coal fields found in the northern France Walloon region, the German Ruhr and Saar areas, as well as Silesia and Ukraine. The industrial revolution followed in its wake.

The favorable position of Great Britain and its lead in both agriculture and industry enabled the country to develop national and international trade quickly. As a result, a large fortune in the form of capital was built up, and new business areas were created through lucrative investments. The concept of monetizability through capital investments was understood early on. Investments resulted in the financing of industrial enterprises, which led to the development of a well-functioning banking system. In the late eighteenth century, banks were founded, and increasingly loans, bills of exchange and bonds developed into a new field of business for traders. Central European countries did not follow to establish joint-stock banks until half a century later. From the past, it is clear that England still plays a dominant role in the financial and banking sectors.

The technological advantages afforded by coal-fired railroad and industrialization consolidated Great Britain's lead on world trade. The aforementioned technological limitations in mining and thus in steel production for rail transport delayed the industrial development of the United States of America and German Reich by 30–40 years. In the meantime, England had become the undisputed center of the world in international capital transfer.

Until that time, Great Britain had not been exposed to any competition, and determined the market price solely through imports of raw materials and exports of products. It was only with the competitiveness of other countries, which had caught up with the market over the following decades, that price regulation through supply and demand began.

The leading sectors in German industrial development were coal mining, iron production and mechanical engineering. The dominant demand and thus the breakthrough of the industrial structures enabled the railroad and the expansion of the railroad network. Around 1840, the length of the railways in the later German Empire was less than 500 km. Ten years later, it was 6000 km and, at the foundation of the Empire in 1871, 19000 km.

The United States, but especially the Northern States, benefited enormously on their first steps toward industrialization through their common history and language and the predominant origin of many immigrants and settlers from Great Britain. Various technologies were also brought and transmitted to the states by British settlers during the industrialization process. Due to the large area of land, it was recognized early on that an expansion of the railroad network would be of crucial importance in order to rapidly advance industrialization. Around 1840, the expansion of the network began drastically in the Northern States and was pushed forward at the same time as Germany.

The southern states, on the other hand, hardly took part in industrialization until 1880. Many regions here had concentrated on the agricultural cultivation of cotton. The sale of raw products played the main role in trade, so that for a long time industrial machines were not necessary. Slavery offered large landowners the opportunity to run their fields at a low cost. The need for efficiency and cost optimization through mechanization was no longer necessary.

Industrialization brought a fundamental change and a reversal of social orders. On the one hand, it brought far-reaching opportunities and a financially rapid rise of the middle class; on the other hand, it brought a loss of position and significance for the nobles and elites due to the changes in modern economic practices. From a political point of view in particular, companies became more prominent and exerted an increasingly strong influence. Improved transportation facilities allowed the population to leave their habitats and settle in urban agglomerations in the industrialized regions. When the population density in industrialized regions increased, this often resulted in desolate working and living conditions. On the other hand, the living conditions such as nutrition, health care and education improved considerably. In its many forms, industrialization was a creeping and peaceful process that gradually changed the world [1].

2.2 The Technological Revolution

For more than two centuries, mankind has benefited from machines. It has helped to multiply the yield of the soil and to transport people over the barriers of space and time. It is not only a mechanical substitute for the human hand, but through the speed of its work, it changed the nature of its products and the lifestyle of generations. Although machines had existed for centuries, the engine provided mechanical power for the first time. Power derived from wind and water was bound to place and opportunity. This problem was overcome by the engine. When James Watt developed the steam engine in 1769, he opened up an inexhaustible source of energy for mankind and thereby triggered the industrial revolution. The lordship over power was once preserved for the most powerful of this world. Horses, slaves, servants and warriors—for thousands of years, influence was driven by animal and human labor [1, 2].

The eighteenth century was initially reserved for the development of the steam engine and not the internal combustion engine. Its rapid development is not a coincidence. Steam can be generated in a heated boiler, and it is then continuously available as long as water is in the boiler. In contrast, fuel, at that time in the form of powder, had to be put back into the cylinder of the engine after each working cycle. The pressures were higher, the control more difficult. This explains why the steam engine was brought to life earlier.

The awareness of having a mechanical force available at all times gradually changed the population's attitude to physical labor. If a process required the performance of heavy physical work, from then on it was an obvious thought to find ways to replace its load with power machines. In addition to agricultural and other activities, as well as textile production, trade and transport, new and industrial professions for the production of machines and other technological products were developed. The production of machines was not the work of a single person. Large enterprises and working communities as well as new forms of life emerged in the industrial boom. It was during this period that the idea of socially linked living and working emerged.

The door to early mobility was opened by the steam locomotive, the beginnings of which were in the1820 s. It was the first means of transport available for everyone that revolutionized the perception of space and time. The history of the American Civil War shows that four decades after the invention of the steam locomotive, railroad played a decisive role in both political and strategic change. The industrial, social and political revolutions since the second half of the eighteenth century were closely related both intellectually and materially.

In the middle of the nineteenth century, the Belgian inventor Etienne Lenoir undertook the development of the two-stroke gas engine. In 1862, the principle was improved by Nicolaus August Otto through the four-stroke process which was named after him. By this time, these new forms of propulsion were large and heavy and were mainly suitable as power engines for industrial purposes. In the middle of 1869, the construction of a factory was started. In Cologne, where the world's first engine factory was built, the center of the large company still stands today. At that time, there were only 40 workers who produced 87 engines under Otto's management in 1869. With the development of the combustion engine, power was available anywhere. It was brought into the hands of the proletariat, and with the revolution of power their rise gradually began [3].

The locomotive only marked the beginning of a new era of mobility. It could not, in the long run, satisfy people's aspiration for freedom as it was tied to rails, stations and railroad networks. In 1878, Amédée Bollée reconsidered the technology of steam power in terms of individual forms of transport and adopted it in the serial production of steam-powered cars. The first models were created by integrating the drives into already existing carriages. The main advantages were that the steam cycle was a closed process and water was available to everyone at all times. Coal and firewood were also cheap fuels that were used to heat the water in the steam boiler. However, the concept also had a major disadvantage: preheating the water until it had reached its operating temperature, which could take up to half an hour before the journey began.

After the steam engine, the electric motor was the second type of mechanical drive that was used as a standard for vehicles. The Flocken electric car, which was developed and presented by Andreas Flocken in Coburg in 1888, was considered the first four-wheeled electric vehicle. The limited range and the long time it took to recharge the battery posed major challenges for this type of drive back then.

At the same time, the first pioneers were working on combustion engines as alternative drive concepts. In 1886, Carl Friedrich Benz applied for a patent for his motor vehicle in Mannheim. In 1889, he presented his Motor Car Number 3 to a broad public at the World Exhibition in Paris. This marked the birth of the internal combustion engine as a standard drive concept, and from then on it took its course. At the same time, Gottlieb Daimler and Wilhelm Maybach presented their version of the first four-wheeled automobile in Stuttgart in 1886. The Daimler-Maybach engine of 1885 was small and light and used a carburetor with fuel injection.

Two decades after the development and series production of the driving concepts—steam, electricity and combustion—it was unclear which of these three would prevail on the market. All of them brought their individual advantages and disadvantages. In 1900, about 4200 cars were built in the USA. Of these, 1600 were steam-powered, 1572 were electrically powered and 1028 had a combustion engine or other drive concept. Only a handful of about 200 manufacturers who had a more or less mature concept to offer survived until 1920. The development of the starter for an internal combustion engine in 1909 was groundbreaking and simplified the cumbersome starting of the engine by means of a hand crank. The higher efficiency and improved driving range, cheap price of oil and the gradual establishment of a network of filling stations were all factors that spoke in favor of the internal combustion engine and gave reason to believe that it would win the race over its competitors [2].

3 Revolution through Simulation for the Development of Powertrains

The computer-aided use of simulation tools for the development of powertrains has become more and more established over the last decades and is indispensable today. For this purpose, simulations are divided into the areas of flow dynamics (3D-CFD), concept simulation (1D/0D), elastohydrodynamics (EHD), kinematics, structure dynamics and reliability (finite element method (FEM) and multi-body simulation (MBS)) and acoustics. Different simulation products are provided for different development phases of powertrains. These mesh with each other throughout the entire development process chain with a strong interaction accordingly. Beginning with the concept phase up to the application of electronic control units, software products contribute their individual strengths and hence allow to benefit from a technologically efficient and targeted application.

Figure 3.1 gives a general overview as to which simulation products are used in the respective phases of powertrain development. In today's modern development units of car manufacturers, development processes are all model-based and in a strong interaction with each other, which makes the use of the complete chain inevitable. The following chapters aim to elaborate the theoretical foundations of the simulation levels and to introduce the fundamental physics behind them.

3.1 Flow Calculation

3D-, 1D- and 0D-CFD tools are generally used to predict the flow path of fluids and to conclude on thermal and thermodynamic behaviors. The selection of a suitable method is made according to the required level of detail. A higher level of detail is always accompanied by a higher computational effort.

The time required for 3D-CFD applications is so high that they are not suitable for overall process calculations. Instead, they are used for the calculation of individual components.

A. Mirfendreski, *Powertrain Development with Artificial Intelligence*,
https://doi.org/10.1007/978-3-662-63863-7_3

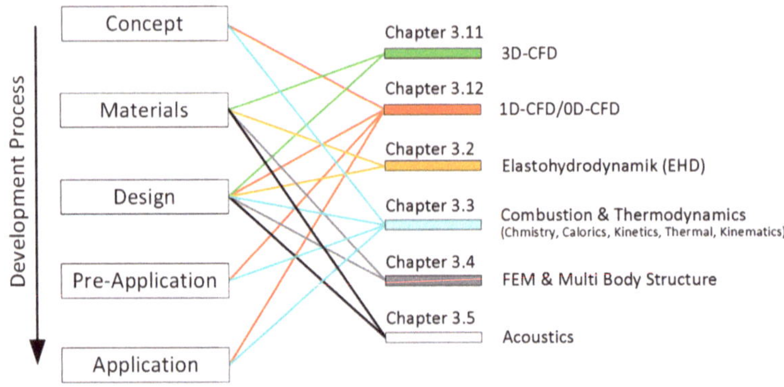

Fig. 3.1 Application areas of simulation for powertrain development

For 1D calculations, the time factor is playing an increasingly important role today. The combination of the increasing performance of computational processors and more efficient algorithms of computational solvers are leading to a new age which allows a complete engine run to be solved within real time. This results in the possibility of coupling simulation tools with real components from the field of electronics. Especially for engine control units, completely new application possibilities open up, which bring significant advantages for the development chain.

A 0D application does not resolve for the location of dependent effects. It predicts identical thermodynamic states for mass, pressure and temperature independent of the spatial coordinate. This level is especially useful if complex overall systems have to be calculated in a short time (vehicle + drive + electronics) or if the effects of a higher resolution are not desired.

3.1.1 3D-CFD

3D-CFD allows gaining differentiated insights into complex flow phenomena. Through its application, fluid-flowing components can be optimized with regard to their design. The following list enumerates some common powertrain areas that are calculated using 3D-CFD:

- Fuel cells: Flow, heat transfer, electrochemical reactions, transport of species in PEM fuel cells,[1] etc.
- Electric drives: thermal management of the electric motor, battery cooling, electrochemical modeling of the battery, aerodynamics, etc.
- Components for the cooling circuit: oil and water circuit (pump, inflow, thermostat, etc.).

[1] PEM: Proton Exchange Membrane.

- Internal combustion engine cylinder intake: airflow, compressor, intercooler (IC), manifold system, port design, valve design (charge movement for tumble and swirl generation), etc.
- Internal combustion engine cylinder: injector position, injection jet (nozzle holes, injection angle, etc.), piston geometry (compression ratio, bowl, cooling channels, etc.).
- Combustion engine cylinder outlet: port design, valve design (charge movement), manifold, EGR path, EGR cooler, turbine inlet and internal flow, VTG, wastegate, aftertreatment, muffler, etc.

The conservation laws for mass, momentum and energy, which are collectively known as the Navier-Stokes equations are the underlying equations forming the basis of numerical fluid mechanics. A flow field is completely characterized by its velocity vector and the state variables pressure, density and temperature as a function of coordinate and time. The **conservation law of mass** is described as follows [4]:

$$\frac{dm}{dt} = 0 \tag{3.1}$$

The equation expresses that the temporal change of mass within a fixed control volume corresponds to the masses added or removed over the boundaries of the volume element. The absolute mass can only increase or decrease if the density of the fluid changes. For incompressible fluids, the conservation equation for the mass in integral form is expressed as follows:

$$\text{Integral}: \frac{\partial}{\partial t}\int_V \rho \mathrm{d}V + \int_S \rho u \cdot n \mathrm{d}S = 0 \tag{3.2}$$

V denotes the control volume, S its surface, n the unit vector perpendicular to S (directed outwards), u the fluid velocity and ρ its density. A corresponding coordinate-free differential form of the continuity equation can be derived by applying the Gauss theorem [4].

$$\text{Vector}: \frac{\partial \rho}{\partial t} + \nabla \cdot \rho u = 0 \tag{3.3}$$

Another, often used representation notation is the Cartesian tensor:

$$\text{Cartesian}: \frac{\partial \rho}{\partial t} + \frac{\partial \rho u_x}{\partial x} + \frac{\partial \rho u_y}{\partial y} + \frac{\partial \rho u_z}{\partial z} = 0 \tag{3.4}$$

Here, (u_x, u_y, u_z) describe the Cartesian components of the velocity vector u. Newton's second equation of motion describes the **conservation of momentum**. It expresses that the momentum of a control volume can be influenced by external forces f [4].

$$\frac{m \cdot u}{\partial t} = \sum f \tag{3.5}$$

The momentum conservation in integral form is described as follows:

$$\text{Integral}: \frac{\partial}{\partial t}\int_V \rho u \mathrm{d}V + \frac{\partial}{\partial t}\int_S \rho u u \cdot n \mathrm{d}S = \int_S T \cdot n \mathrm{d}S + \int_V \rho b \mathrm{d}V \tag{3.6}$$

The left side of the equation describes the momentum of a fluid within a control volume. Its change over time corresponds to the sum of acting surface forces (shear and normal forces) and external forces. These may consist of gravitational, centrifugal, Coriolis or electromagnetic forces. These are summarized in the variable b [4].

The stress tensor T for a Newtonian fluid describes the molecular transport rate of a momentum and can be represented in the Cartesian coordinate system as follows [4]:

$$T_{ij} = -\left(p + \frac{2}{3}\mu \frac{\partial u_j}{\partial x_i}\right)\delta_{ij} + 2\mu D_{ij} \tag{3.7}$$

D represents the tensor of the deformation rate, δ_{ij} the Kronecker symbol ($\delta_{ij} = 1$ for $i = j$, $\delta_{ij} = 0$ for $i \neq j$), and p and μ represent the pressure and the dynamic viscosity of the fluid, respectively [4].

For the description of the viscous part of the stress tensor, the following notation is also used in the literature:

$$\tau_{ij} = 2\mu D_{ij} - \frac{2}{3}\mu \delta_{ij} \nabla u \tag{3.8}$$

with

$$D_{ij} = \frac{1}{2}\left(\frac{\partial u_i}{\partial x_j} + \frac{\partial u_j}{\partial x_i}\right) \tag{3.9}$$

The coordinate-free vector representation of the conservation of momentum is

$$\text{Vector}: \nabla u_i \rho u = u_i \nabla \cdot (\rho u) + \rho u \cdot \nabla u_i \tag{3.10}$$

If the viscous part of the stress tensor τ_{ij} is used in Eq. 3.6 and gravity is considered the only external force, the shape is obtained in Cartesian tensor notation:

$$\text{Cartesian}: \frac{\partial \rho u_i}{\partial t} + \frac{\partial \rho u_j u_i}{\partial x_j} + \frac{\partial \tau_{ij}}{\partial x_j} - \frac{\partial p}{\partial x_i} + \rho g_i = 0 \tag{3.11}$$

The **energy conservation equation** can be represented in different forms, depending on which physical quantity is considered as variable (temperature, internal energy, thermal enthalpy, total enthalpy, etc.). A common representation for a flow can be described using the enthalpy h and the thermal conductivity k [4]:

$$\text{Integral}: \frac{\partial}{\partial t}\int_V \rho h \mathrm{d}V + \int_S \rho h u \cdot n \mathrm{d}S = \int_S k\nabla T \cdot n \mathrm{d}S + \int_V (u \cdot \nabla p + S' \cdot \nabla u)\mathrm{d}V + \frac{\partial}{\partial t}\int_V p \mathrm{d}V \tag{3.12}$$

S' corresponds to the viscous part of the stress tensor. For the coordinate-free vector representation of the conservation of energy theorem, the following results:

$$\text{Vector}: \frac{\partial(\rho\phi)}{\partial t} + \nabla \cdot (\rho\phi u) = \nabla \cdot (\Gamma\nabla\rho) + q_\phi \tag{3.13}$$

For the conservation of a scalar, ϕ represents the amount of the scalar per unit mass, e.g. the specific enthalpy or the internal energy. The differential form of the generic conservation equation in Cartesian coordinates and tensor notation is

$$\text{Cartesian}: \frac{\partial(\rho\phi)}{\partial t} + \frac{\partial(\rho u_i\phi)}{\partial x_j} = \frac{\partial}{\partial x_j}\left(\Gamma\frac{\partial\phi}{x_j}\right) + q_\phi \tag{3.14}$$

The variable Γ here describes the diffusion coefficient for the quantity ϕ, and q_ϕ designates sources and sinks of ϕ, which are supplied or discharged to the control system. By the conservation laws for mass, momentum and energy, a system is completely described together with the thermal equation state for ideal gases and the relations between the specific gas constant R, the isentropic exponent κ and the specific heat capacities c_p as well as c_v [5]:

$$p \cdot v = m \cdot R \cdot T, \kappa = \frac{c_p}{c_v}, R = c_p - c_v \tag{3.15}$$

All in all, a system of equations consisting of nonlinear, partial differential equations (PDE) results. For a 3D-CFD calculation process, these are iteratively calculated at each point of the discretized structure (mesh) at each individual time step.

3.1.2 1D-CFD

If a complete engine process including an existing air path system is the objective of investigation, a 1D approach is a good choice. In a 1D simulation, the fluid flow is considered exclusively along the main flow direction, which results in a moderate computational effort in relation to the 3D-CFD. In addition to the calculation of an air path, 1D flow models using the example of an engine are suitable for the calculation of charge exchange, exhaust gas turbocharger design, geometry optimization, and the design and dimensioning of cooling and oil circuits.

1D models are also based on the conservation equations of mass, momentum and energy. As a result of the reduced dimensions, a 1D flow calculation can be expressed in a much simpler way compared to 3D-CFD. The following basic equations are presented in [5].

With regard to **mass conservation**, the following simplification is made for the 1D, starting from the 3D, view, by omitting the local coordinate resolution in y- and z-directions:

$$\frac{\partial}{\partial t} + \frac{\partial m_x}{\partial x} + \cancelto{0}{\frac{\partial m_y}{\partial y}} + \cancelto{0}{\frac{\partial m_z}{\partial z}} = 0 \tag{3.16}$$

$$\rightarrow \frac{\partial}{\partial t} + \frac{\partial m_x}{\partial x} = 0 \tag{3.17}$$

By applying the continuity equation, the mass can be replaced. This results in the vectorial notation:

$$\frac{\partial}{\partial t} + \frac{\partial m_x}{\partial x} = \frac{\partial}{\partial t} + \frac{\partial(\rho_x u_x A_x)}{\partial x} \tag{3.18}$$

A_x corresponds to the flow area of a pipe transverse to the flow direction. Due to the 1D approach, the x-component of the flow velocity corresponds to the entire flow vector $u_x = u$. Hence, the conservation of mass for 1D leads to

$$\frac{\partial \rho}{\partial t} + \rho \frac{\partial u}{\partial x} + u \frac{\partial \rho}{\partial x} + \frac{\rho u}{A} \frac{\partial A}{\partial x} = 0 \tag{3.19}$$

For the **conservation of momentum**, the reduction of the single terms below can be performed based on the 3D view. The surface forces and the external forces are combined and represented by a friction coefficient f_R:

$$\frac{\partial \rho u_i}{\partial t} + \frac{\partial \rho u_j u_i}{\partial x_j} + \frac{\partial \tau_{ij}}{\partial x_j} - \frac{\partial p}{\partial x_i} + \rho g_i = 0 \tag{3.20}$$

$$\rightarrow \frac{\partial \rho u}{\partial t} + u \frac{\partial u}{\partial x} + \frac{\partial p}{\partial x} + f_R = 0 \tag{3.21}$$

If one follows the same scheme and tries to get from the 3D view of **energy conservation** to the 1D one, Eq. 3.22 can take different forms depending on which scalar is chosen. By inserting the equations for continuity from Eq. 3.17 and the conservation of momentum from Eq. 3.21, the energy equation can be represented in a generic form as a function of the total pressure differential, see Eq. 3.23, [5, 6]:

$$\frac{\partial(\rho \phi)}{\partial t} + \frac{\partial(\rho u_i \phi)}{\partial x_j} = \frac{\partial}{\partial x_j}\left(\Gamma \frac{\partial \phi}{x_j}\right) + q_\phi \tag{3.22}$$

$$\rightarrow \frac{\partial p}{\partial t} + u \frac{\partial p}{\partial x} + (q_\phi + u \cdot f_R)\rho = 0 \tag{3.23}$$

3.1.3 0D-CFD

A much simpler description of an air path is based on zero-dimensional (0D) models. By decoupling a system from its path and time, the flow process is detached from a path and can be represented solely by a time dependency. In a considered subsystem, the thermodynamic state is spatially constant at any discrete time.

Even in a 1D flow modeling, parts of the air path in the form of air volumes can be considered dimensionless. In this case, the reflection is calculated with a characteristic and representative length of a volume. Additionally, other subsystems of the engine as well as the combustion chamber, exhaust turbocharger, etc. are modeled on the basis of a 0D approach.

Filling and emptying models

The filling and emptying method is a concept commonly used in practice for calculating fluid-flow pipe systems. Here, pipelines of the air path system are combined to spherical containers with corresponding volumes, which is why this type of model is also called "container model". These are separated from each other by means of orifices or valves with fixed or variable cross-sections. Changes of state within a container are calculated considering transient filling and emptying processes. A positive volume flow between adjacent containers is generated by filling the front container and emptying the rear one.

A crucial assumption is that pressure and temperature within the containers are balanced without delay. For each discrete calculation step, a complete mixing of the tank contents takes place. The flow thus moves through the air path system at an infinitely high speed of sound.

For a calculation of transient processes, it is assumed that the flow can be treated stationary for small time intervals, which leads to deviations in the result. Gas dynamic flow effects cannot be reproduced with an R&D method, since the latter is not suitable for the investigation of concepts such as resonance charging at the cylinder inlet and shock charging at the exhaust turbocharger [5].

A system is described by the average values of pressure, temperature, mass, internal energy and the heat flow across the system boundaries.

For the calculation of container models, the conservation laws for mass and energy are calculated, the consideration of the conservation of momentum is omitted due to the absence of a local resolution. Thus, the system is greatly simplified compared to a 1D approach. The **conservation of mass** leads to the following differential form from a simplified 3D consideration:

$$\frac{\partial \rho}{\partial t} + \cancelto{0}{\frac{\partial \rho u_x}{\partial x}} + \cancelto{0}{\frac{\partial \rho u_y}{\partial y}} + \cancelto{0}{\frac{\partial \rho u_z}{\partial z}} = 0 \tag{3.24}$$

$$\rightarrow \frac{\partial \rho}{\partial t} = 0 \tag{3.25}$$

From this, a temporally solvable representation can be derived:

$$\frac{\mathrm{d}}{\mathrm{d}t}m(t) = \dot{m}_{\text{in}}(t) - \dot{m}_{\text{out}}(t) = 0 \tag{3.26}$$

The **conservation law for energy**, based on the 3D view, can be reduced accordingly to the following form:

$$\frac{\partial(\rho\phi)}{\partial t} + \cancelto{0}{\frac{\partial(\rho u_j \phi)}{\partial x_j}} = \cancelto{0}{\frac{\partial}{\partial x_j}\left(\Gamma\frac{\partial\phi}{x_j}\right)} + q_\phi \tag{3.27}$$

$$\rightarrow \frac{\partial\rho\phi}{\partial t} = q_\phi \tag{3.28}$$

If, for example, for the scalar ϕ the internal energy of a system is considered, the temporal change takes the following form:

$$\frac{\mathrm{d}U}{\mathrm{d}t} = \frac{\mathrm{d}Q_{\mathrm{W}}}{\mathrm{d}t} + \frac{\mathrm{d}H_{\mathrm{zu}}}{\mathrm{d}t} - \frac{\mathrm{d}H_{\mathrm{ab}}}{\mathrm{d}t} \tag{3.29}$$

Using the thermal equation of state for ideal gases, the internal energy can also be expressed as a function of pressure, temperature and the isentropic exponent κ:

$$\frac{\partial U}{\partial t} = \frac{\partial}{\partial t} \cdot \frac{p \cdot V}{\kappa - 1} = \frac{1}{\kappa - 1} \cdot \frac{\partial p}{\partial t} \cdot V + \frac{1}{\kappa - 1} \cdot p \cdot \frac{\partial V}{\partial t} \tag{3.30}$$

Together with the caloric correlations

$$U = m \cdot c_v \cdot T, H = m \cdot c_p \cdot T \tag{3.31}$$

and Eqs. 3.29 and 3.31 while disregarding the heat flows over the walls result in the following derivations for p and T:

$$\frac{\mathrm{d}T}{\mathrm{d}t} = \frac{T \cdot R}{c_v \cdot p \cdot V}\left[\left(\dot{H}_{\mathrm{zu}}(t) - \dot{H}_{\mathrm{ab}}(t)\right)\left(1 - \frac{c_v}{c_p}\right)\right] \tag{3.32}$$

$$\frac{\mathrm{d}p}{\mathrm{d}t} = \frac{\kappa \cdot R}{V \cdot c_p}\left[\dot{H}_{\mathrm{zu}}(t) - \dot{H}_{\mathrm{ab}}(t)\right] \tag{3.33}$$

To determine the mass flow over the throttling points (see Fig. 3.2), a flow equation according to *de Saint-Venant* is used for stationary, adiabatic flows. A detailed derivation can be read in [5]:

$$\dot{m} = \alpha_K \cdot A_{\mathrm{zu}} \cdot \sqrt{p_0 \cdot \rho_0} \cdot \psi, \psi = \sqrt{\frac{2\kappa}{\kappa - 1}\left(\left(\frac{p_1}{p_2}\right)^{\frac{2}{\kappa}} - \left(\frac{p_1}{p_2}\right)^{\frac{\kappa+1}{\kappa}}\right)} \tag{3.34}$$

Here, A_{in} is the geometric cross-section, p_0 and ρ_0 the pressure and density in the tank in front of the orifice, α_{K} the flow coefficient and p_1 the pressure in the cross-section. The term ψ is also called outflow function.

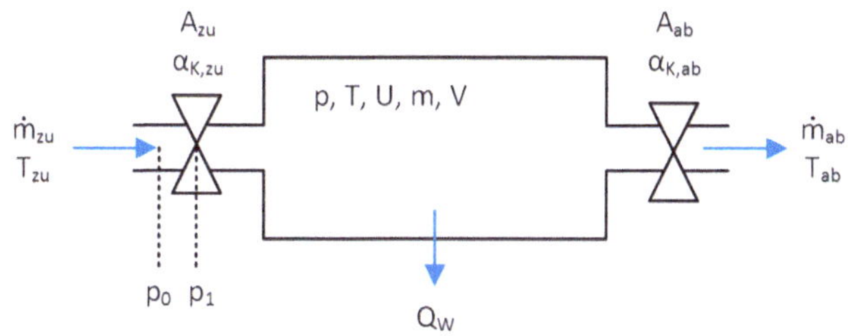

Fig. 3.2 State variables, fluid and energy flow of a container model

By means of the flow coefficients, real effects on the flow through a valve, such as constriction, are taken into account, whereby the actual flow cross-section is smaller than the geometric one. This coefficient describes the ratio of the actual mass flow to the theoretically possible mass flow.

The maximum of the outflow function results from an extreme value determination. At the point ψ_{max}, the critical pressure ratio $\left(\frac{p_1}{p_0}\right)_{krit}$ is just reached, as the following equation results:

$$\frac{\partial \psi}{\partial \left(\frac{p_1}{p_0}\right)} = 0, \left(\frac{p_1}{p_0}\right)_{krit} = \left(\frac{2}{\kappa + 1}\right)^{\frac{\kappa}{\kappa - 1}} \tag{3.35}$$

As can be seen in Fig. 3.3, the value of the function increases up to the critical pressure ratio. At this point, the speed of sound and thus the maximum possible mass flow is reached in the narrowest cross-section. To the left, the mass flow can be increased by increasing p_0 or decreasing p_1 [5, 7].

Fig. 3.3 Flow function as a function of isentropic exponent and pressure ratio [5]

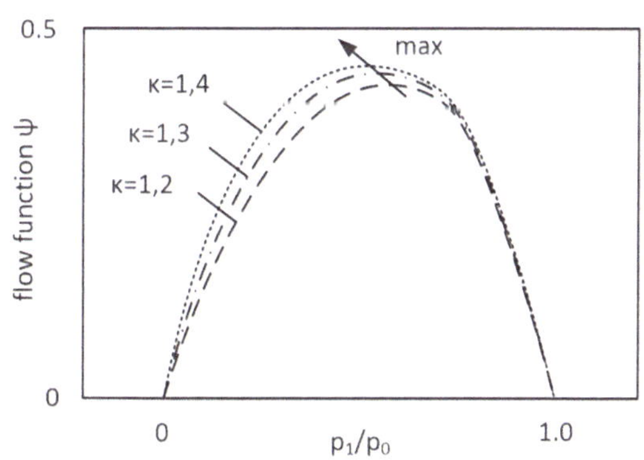

Mean value models

A mean value model (MVM) is a subordinate and compressed form of the filling and emptying models. The difference lies in the fact that, in contrast to a conventional engine process calculation, the working process is not crank angle-resolved but considered averaged over a cycle.

Such models are used when a crank angle-resolved view of the engine can be dispensed with, or when even shorter computing times are required in contrast to the filling and emptying method. This can be the case, for example, for fast parameter studies, integration of the engine model into a complete vehicle environment (longitudinal dynamics) or for the coupling of electronic control units, where real-time capability is mandatory.

3.2 Elastohydrodynamics (EHD) and Tribology

Tribology deals with the science of interactions of surfaces in relative motion, which includes the principles of friction, lubrication and wear. Its application is interdisciplinary and is based on physics, chemistry, materials science and engineering (Fig. 3.4).

One of the most important applications of tribology in the field of engineering can be found in the design and layout of bearings. In sliding roller contacts, such as rolling bearings, gears and cam followers, the treads are elastically deformed due to high contact pressures. For plastics such as seals, deformation can even occur at low pressure. This results in a much larger surface area for carrying the load force than would be the case without deformation. During operation, material stresses can hence be kept within specified limits despite high forces [8].

Lubrication under consideration of deformations has been increasingly understood in the last decades as a describable phenomenon and transferred into mathematical models. This has led to the development of simulation tools that help to carry out sensitivity studies, influence processes, understand effects more precisely and ultimately transfer studied knowledge to implement it into the design.

Elastohydrodynamic lubrication is generally referred to when the elastic deformation of two contact bodies is equal to or approximately greater than the thickness of the lubricant film. If the deformation is negligible, on the other hand, the lubrication is referred to as purely hydrodynamic. If the contact bodies are stiff, as is the case with metals, this is referred to as hard EHD lubrication. Soft EHD lubrication, on the other hand, occurs when one or both contact bodies have a low modulus of elasticity. In the case of endurance runs, it is a matter of stationary EHD lubrication. If, on the other hand, a time-discrete behavior is of interest, such as during a warm-up or dynamic load on contact bodies, this is referred to as transiently loaded EHD lubrication.

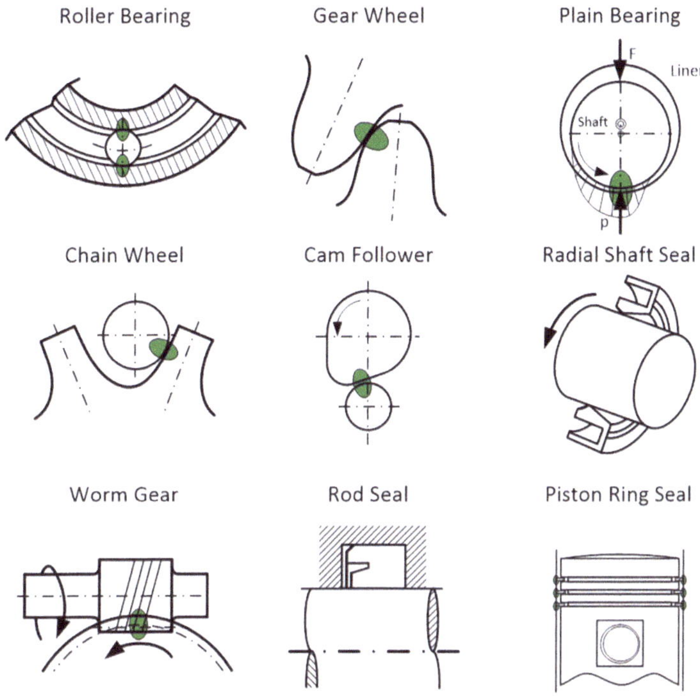

Fig. 3.4 Examples of EHD-lubricated contacts [8]

In general, a contact between any two bodies locally forms an elliptical pressure surface. In one extreme case, the load state can reach a circular contact, i.e. a point contact, or approximately a bath-shaped contact or line contact in the other extreme case. In a line contact, the bodies are in a plane deformation state.

To describe a tribological system, it is necessary to know important parameters such as lubricant film thickness and pressure, load capacity, friction and material stresses. Problems of the EHD can generally be divided into stationary hard EHD, stationary soft EHD, thermal EHD and transient EHD.

The course of the friction force as a function of the friction velocity in the case of hydrodynamic friction is typically described by the so-called Stribeck curve (see Fig. 3.5). This curve gives insight into the effect of the lubricant film thickness on the friction coefficient as a function of oil viscosity, relative speed of the material pairings and the normal force.

If there is no relative movement between two pairs of materials, this is called dry friction. If a force greater than the static friction force is applied, a relative movement is initiated. In the first phase, the lubricant separates the material pairings from each other only at the molecular level—this phase is called boundary lubrication.

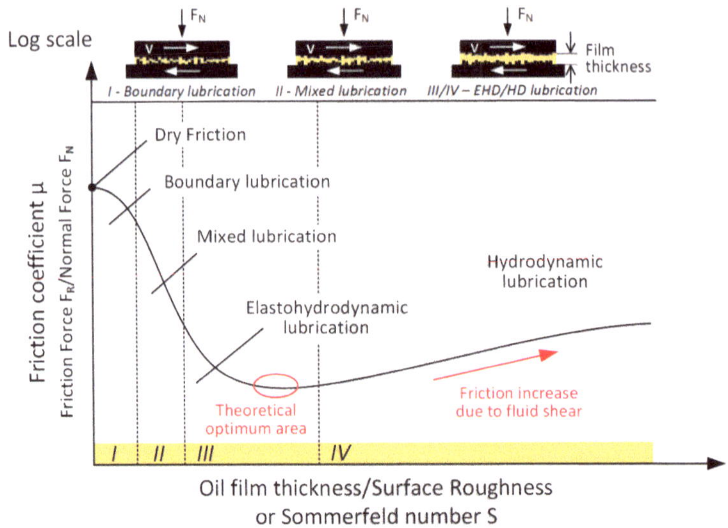

Fig. 3.5 Stribeck curve: Effect of the lubricant film on the friction coefficient of two materials [9]

If a thin lubricating film forms between the bodies so that they are only separated by slight material roughness, the bodies can slide off on the lubricant—this phase is known as mixed lubrication.

When mixed friction is overcome, the material pairs are completely separated by the lubricating film in the next phase. This phase, which is called liquid friction, can be divided into an elastohydrodynamic and a hydrodynamic phase, where material wear is lowest. As more and more layers of the lubricant slide on top of each other with increasing relative speed, the frictional force increases again in the hydrodynamic phase. In general, it is desirable to achieve a coefficient of friction in the range of the minimum for any design of material pairings and oil [8].

In its simplest form, the entire problem of elastic hydrodynamic lubrication (EHL) is described by five equations as follows.

1. The Reynolds equation describes the flow of a Newtonian fluid in a narrow gap.

It was derived from the Navier-Stokes equation to express the character of a slow and viscous flow. Inertial forces as well as external forces versus viscous forces are neglected for simplification. A second simplification is done based on the geometrical conditions of a narrow gap: The dimensions in z-direction are much smaller than in x- and y-directions, see Fig. 3.6. Applying the condition of slip resistance along the wall, the velocity profile is

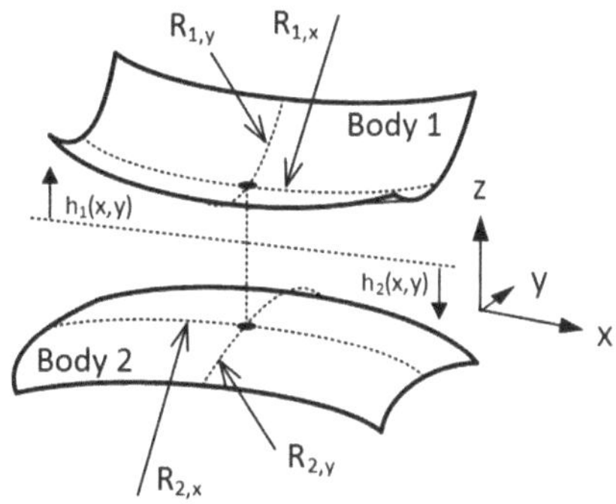

Fig. 3.6 Coordinate system in the contact point of two bodies

obtained as a function of the z-coordinate. The continuity of the mass flow in a narrow gap, extending from one surface to the other, results as follows:

$$\frac{\partial}{\partial x}\left(\frac{\rho h^3}{12\eta}\frac{\partial p}{\partial x}\right)\frac{\partial}{\partial y}\left(\frac{\rho h^3}{12\eta}\frac{\partial p}{\partial y}\right)-\frac{\partial\left(u_m\rho h\right)}{\partial x}-\frac{\partial(\rho h)}{\partial t}=0 \tag{3.36}$$

In the case of a line contact, the contact dimensions in the y-direction are very large compared to those in the x-direction, so that the following simplified equation applies:

$$\frac{\partial}{\partial x}\left(\frac{\rho h^3}{12\eta}\frac{\partial p}{\partial x}\right)-u_m\frac{\partial h}{\partial x}=0 \tag{3.37}$$

Here p is the pressure, h the film thickness, u_m the average surface speed, η the viscosity and ρ the density of the oil.

2. The elastic deformation equation describes the influence on the deformation of the oil gap.

To approximate the elastic deformations of real bodies, two assumptions are made: The deformation is linearly elastic and the two contact bodies have uniform and isotropic properties. The contact dimensions a are small compared to the dimensions of the body ($a << R_x$), so that the additional assumption of an approximation of the bodies by two semi-infinite half-spaces can be made.

Both hypotheses are generally valid and the obtained approximate values agree very well with experimental results. Thereby, the elastic deformation $h(x, y)$ according the pressure distribution $p(x, y)$ can be approximated by

$$h(x,y)=h_0+\frac{x^2}{2R_x}+\frac{y^2}{2R_y}+\frac{2}{\pi E'}\iint_{-\infty}^{\infty}\frac{p(x',y')\mathrm{d}x'\mathrm{d}y'}{\sqrt{(x-x')^2+(y-y')^2}} \tag{3.38}$$

For a line contact with a large contact length in the y-direction compared to that one in the x-direction, the equation can be simplified to

$$h(x) = h_0 + \frac{x^2}{2R_x} + \frac{2}{\pi E'} \iint_{-\infty}^{\infty} p(x') \ln \left(\frac{x - x'}{x_0} \right)^2 \mathrm{d}x' \tag{3.39}$$

3. The viscosity-pressure relationship describes viscosity as a function of pressure. The simplest form is known under the exponential approach of Barus:

$$\eta(p) = \eta_0 \cdot e^{(\alpha \cdot p)} \tag{3.40}$$

Here, η_0 is referred to as the atmospheric viscosity and α as the pressure viscosity coefficient.

4. The density-pressure relationship expresses the density as a function of pressure according to the approach of Dowson and Higginson:

$$\rho(p) = \rho_0 \frac{5.9 \cdot 10^8 + 1.34p}{5.9 \cdot 10^8 + p} \tag{3.41}$$

5. The principle of interaction means that the integral of the pressure distribution from the Reynolds equations must balance the externally applied load w in order to establish a balance of forces. For the 2D problem, this condition is

$$w = \iint_{-\infty}^{\infty} p(x', y') \, \mathrm{d}x' \mathrm{d}y' \tag{3.42}$$

w stands for the applied load in the 2D case. For the one-dimensional load case (line contact), the applied load is

$$w = \iint_{-\infty}^{\infty} p(x') \, \mathrm{d}x' \tag{3.43}$$

For both the 1D and the 2D problem, the local equations for describing pressure and film thickness distributions $p(x, y)$ and $h(x, y)$ are supplemented by the global Eqs. 3.42 and 3.43, so that the film thickness h_0 can be determined using the Eqs. 3.38 and 3.39 [10].

3.3 Combustion Chamber (Combustion and Thermodynamics)

The Science of Combustion is a discipline that deals with the physical and chemical processes of combustion and the different characteristics of flames. Its complexity results from the contact and interaction of many other fields of physics. Examples mentioned are fluid dynamics, thermodynamics, calorics, reaction kinetics, mechanics and more. If the aim is to simulate combustion, a chain of consecutive calculations and thus boundary conditions must be well predicted in order to make further reliable calculations.

Figure 3.7 illustrated the chamber of a cylinder. It is geometrically limited by the cylinder walls to the sides, the piston to the bottom and the valves (cylinder head) to the top. The complexity of the combustion process from the viewpoint of multiple physical and chemical processes all taking place in the cylinder gets very clear from this picture.

According to the law of conservation of mass, the change of mass in a control volume is always constant, which corresponds to the sum of all mass flows entering and leaving the volume:

$$dm = dm_{in} - dm_{out} + dm_{Fuel} - dm_{Blowby} \tag{3.44}$$

In their entirety, all the processes mentioned generate incoming and outgoing energy flows that can be summarized in an overall energy balance according to the first law of thermodynamics. Putting them into equilibrium results in the following differential function over time:

$$dU = dH + dQ_B - dQ_W - dW \tag{3.45}$$

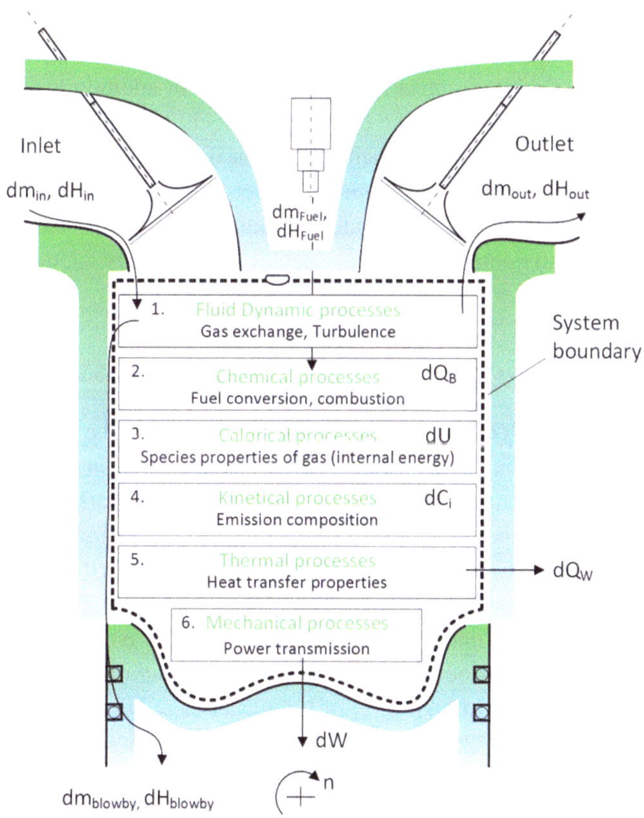

Fig. 3.7 Physical/chemical process sequences in a cylinder

U stands for the internal energy within the system boundaries, H for the supplied and removed enthalpy flows, Q_B for the released combustion energy, Q_W for the heat losses and W for the mechanical power output. If the enthalpy flow in the system boundary "combustion chamber" shown in the figure are resolved, the following results are obtained:

$$dU = dH_{in} - dH_{out} - dH_{blowby} + dH_{Fuel} + dQ_B - dQ_W - dW \tag{3.46}$$

For the temporal change of the internal energy during a working cycle, usually the differential from U is solved according to the crank angle, which results in the following term:

$$\frac{dU}{d\varphi} = h_{in}\frac{dm_{in}}{d\varphi} - h_{out}\frac{m_{out}}{d\varphi} - h_{out}\frac{dm_{Blowby}}{d\varphi} + h_{Fuel}\frac{dm_{Fuel}}{d\varphi} + \frac{dQ_{Fuel}}{d\varphi} - \frac{dQ_W}{d\varphi} - \frac{dW}{d\varphi} \tag{3.47}$$

In the following sections, all six physical processes are discussed in sequence according to Fig. 3.7 and their theoretical foundations are presented.

1. Fluid dynamics (dH): During the charge exchange, the calculation basis of fluid dynamics is used to determine the mass flow of fresh air (in case of direct injection) or fuel-air mixture (in the case of mixture intake) into the cylinder. In addition to the pressure gradient between the inlet channel and the cylinder chamber, it is crucial that the pulsations of the inlet pressure are composed in such a way that there is a pressure peak during the opening phase of the inlet valve, thus promoting better filling. The basics of fluid dynamics for this were presented in Chap. 3.1. A change of enthalpy dH that is fed to or discharged from a system is basically obtained from the specific heat capacity c_p, the mass m and the temperature difference $dT = (T_{out} - T_{in})$ [5].

$$dH = c_p \cdot m \cdot dT \tag{3.48}$$

2. Chemical processes (dQ_B): The introduction of a fuel into the combustion chamber and the conversion of its bound chemical energy into heat follows in this process step. The ignition of the fuel-air mixture can be effected externally by a spark plug (petrol engine) or self-ignition (diesel engine) or by one of many alternative combustion processes such as compressed auto ignition (CAI) or homogeneous charge compressed ignition (HCCI).

The burning speed and thus the time needed to convert the chemical energy of the fuel are very decisive. Typically, one tries to make it as quick as possible, since the highest thermodynamic efficiency can be achieved with an early center of combustion ($\sim 6° - 8°$ crank angle). To achieve this, a high level of turbulence is required by the end of the gas exchange phase.

Engine power, efficiency and emissions are controlled by the combustion of the fuel-air mixture. In order to understand the engine operation in its entirety, many individual building blocks are required for relevant combustion phenomena. The phenomena differ for the main engine type spark ignition and diesel. In gasoline engines, the fuel is either mixed with air in the engine intake system or injected directly into the cylinder. After compression of the fuel-air mixture, an electrical discharge of the spark plug triggers the combustion process. The ignition spark develops into a flame which spreads into the cylinder space until it reaches the walls of the combustion chamber on the sides or the piston downwards and extinguishes. An undesirable phenomenon that occurs is the spontaneous or uncontrolled self-ignition of a significant portion of the fuel-air mass before it is ignited by the flame. The explosive phenomenon is the cause of engine knocking, which can lead to engine damage due to the high pressures generated.

In diesel engines, the fuel is injected into the cylinder at high pressure and high temperature. The self-ignition of parts of the developing mixture of already injected and vaporized fuel with the hot air initiates the combustion process. Unlike the gasoline engine, this process is turbulent. The composition of the fuel-air mixture therefore plays a decisive role in diesel combustion.

The composition of fuel and air is called stoichiometry and describes the reactants of a combustible mixture and the composition of the products. Since the relationships depend on the conservation of the mass of each chemical element of the reactants, the relative compositions of the fuel and the fuel/air ratio are used as the basis for calculation. If sufficient oxygen is available, a hydrocarbon fuel can completely oxidize. The carbon in the fuel is then converted into carbon dioxide CO and the hydrogen H_2 into water H_2O [5].

For a fuel with the composition $C_xH_yS_qO_z$, the following reaction takes place in the case of complete conversion:

$$C_xH_yS_qO_z + \left(x + \frac{1}{4}y + q - \frac{1}{2}z\right) \cdot O_2 = xCO_2 + \frac{y}{2} \cdot H_2O + qSO_2 \tag{3.49}$$

with the stoichiometric coefficients and the mass fractions of the elements carbon c, hydrogen h, sulfur s and oxygen o contained in the fuel:

$$x = \frac{M_B}{M_C}c,\ y = \frac{M_B}{M_H}h,\ q = \frac{M_B}{M_S}s,\ z = \frac{M_B}{M_O}o \tag{3.50}$$

The stoichiometric air requirement leads to

$$\begin{aligned} L_{st} &= \frac{1}{x_{O_{2},L}} \cdot \frac{m_{O_{2,st}}}{m_B} = \frac{1}{x_{O_{2},L}}\left(\frac{M_{O_2}}{M_C} \cdot \frac{1}{4}\frac{M_{O_2}}{M_H} \cdot h + \frac{M_{O_2}}{M_S} \cdot s - o\right) \\ &= \frac{1}{0.232} \cdot (2.664 \cdot c + 7.937 \cdot h + 0.998 \cdot s - o) \end{aligned} \tag{3.51}$$

For an engine combustion, the ratio of actual air mass m_L is related to the stoichiometric air mass $m_{L,st}$ and expressed the air ratio λ:

$$\lambda = \frac{m_L}{m_{L,st}} = \frac{m_L}{m_B} \cdot L_{st} \tag{3.52}$$

The mixing heating value H_G of air and fuel is composed of the fuel mass m_B. In case of external mixture formation (naturally aspirated engine), the lower fuel heating value H_u and the mixture volume V_G are applied in the following equation:

$$H_G = \frac{m_B \cdot H_u}{V_G} = \frac{\rho_G \cdot H_u}{\lambda \cdot L_{st} + 1} \tag{3.53}$$

For engines with internal mixture formation (diesel engines, DI gasoline engines), the fuel mass and the lower heating value are related to the air volume V_L:

$$H_G = \frac{m_B \cdot H_u}{V_L} = \frac{\rho_L H_u}{\lambda L_{st}} \tag{3.54}$$

3. Caloric processes (dU): After the flow-relevant processes of the gas exchange and the chemical processes of the fuel have been discussed, the next sequence according to Fig. 3.7 is followed by the determination of the caloric content. This deals with the material properties of the working gas and is ultimately used to determine the internal energy U within the system boundaries of the combustion chamber and to calculate the term dU from the energy balance according to Eq. 3.45 [5].

A possible approach to the calculation of the internal energy U lies in the description of the individual components of a gas mixture and can be described as

$$dU = \sum_i m_i du_i = \sum_i m_i d(h_i - RT_i) \tag{3.55}$$

The best known approaches to describing the specific internal energy of flue gas, i.e. the combustion products, were developed by Justi in 1938 and Zach-arias in 1968 [11, 12]. Neglecting dissociation effects, i.e. backward reactions of already formed species, and the pressure dependence of reactions, Justi presented a polynomial approach, which represents the specific internal energy as a function of temperature and air ratio. Zacharias' approach made it possible to determine the flue gas more precisely by breaking it down into its components, an approach that could only be overcome with the use of the first mainframe computers.

The internal energy of each gas component can be calculated separately, since its standard enthalpies of formation, reaction enthalpies and the respective molar heat are given in tables as in NIST JANAF [13]. By knowing the respective proportions of these individual components, the total internal energy of a gas can be calculated.

Typical gases emitted by an internal combustion engine such as oxygen, nitrogen, fuel vapor, carbon dioxide and water vapor may be considered ideal, described by the following equation:

$$pV = mRT = n\tilde{R}T \tag{3.56}$$

where p is pressure, V volume, m gas mass, $\tilde{R}$ universal gas constant, T temperature and n molar mass.

Considering the cylinder charge, an ideal gas, the specific heat capacity, and enthalpy and entropy for any reaction can be determined according to the following formulas:

$$\frac{\tilde{c}}{\tilde{R}} = a_1 + a_2T + a_3T^2 + a_4T^3 + a_5T^4 \tag{3.57}$$

$$\frac{\tilde{h}}{\tilde{R}T} = a_1 + \frac{a_2}{2}T + \frac{a_3}{3}T^2 + \frac{a_4}{4}T^3 + \frac{a_5}{6}T^4 + \frac{a_6}{T} \tag{3.58}$$

$$\frac{\tilde{s}}{\tilde{R}} = a_1 \cdot lnT + a_2T + \frac{a_3}{2}T^2 + \frac{a_4}{3}T^3 + \frac{a_5}{4}T^4 + a_7 \tag{3.59}$$

4. Kinetic processes (dC_i): Kinetics is a subarea of physical chemistry and deals with the time sequence of chemical reactions or physical-chemical processes. It can generally be divided into the subareas of micro- and macrokinetics. While microkinetics is mainly concerned with the time sequence of chemical reactions and their mathematical description, macrokinetics takes global influences of thermodynamics as well as heat and mass transport processes (diffusion) into account.

In the application for the internal combustion engine and the area of simulation, reaction kinetics is particularly relevant, since it ensures the calculation basis for many fundamental processes. These include influences of different fuel types on the chemical processes in the combustion chamber (including combustion speed and knock resistance), the generation of raw emissions and catalytic processes of exhaust gas aftertreatment [5].

Reaction speed

A chemical reaction with the educts A and B, which form the products C and D, is described according to the following formalism:

$$\alpha \mathrm{A} + \beta \mathrm{B} + \gamma \mathrm{C} \underset{k_{1,r}}{\overset{k_{1,v}}{\rightleftharpoons}} \delta \mathrm{D} + \varepsilon \mathrm{D} \tag{3.60}$$

α, β, γ, δ and ε represent the stoichiometric coefficients of the reaction, and k_f and k_r describe the reaction rate of the forward and backward reactions, respectively. The temporal change of a species concentration (taking component A as an example) can be determined for the given chemical reaction equation with the following empirical approach:

$$\frac{dC_A}{dt} = \alpha \left(k_f[\mathrm{A}]^\alpha \cdot [\mathrm{B}]^\beta \cdot [\mathrm{C}]^\gamma - k_r[\mathrm{D}]^\delta \cdot [\mathrm{E}]^\varepsilon\right) \tag{3.61}$$

The reaction rate k_f and k_r are experimentally determined quantities, which are summarized in extensive tables for all chemical reactions. The measurements originate from experiments in shock wave reactors or similar. Since they are also strongly temperature-dependent, they are corrected using an Arrhenius approach:

$$k = A \cdot T^b \cdot e^{-\frac{E_A}{R \cdot T}} \tag{3.62}$$

In particular, the speed of the reaction depends on the activation energy E_A, which must be applied to initiate the reaction, and the absolute temperature T. For very high temperatures, the term converges against the pre-exponential factor AT^b. In this case, the velocity is described by the shock kinetics of the molecules. The constant A, the temperature coefficient b and the activation energy E_A can also be taken from tables.

Reaction equilibrium

At the molecular level, a chemical reaction always takes place in two directions. From the difference between the forward and backward reactions, it is possible to draw conclusions about the direction of the reaction. The chemical equilibrium is therefore only a special case in which the forward and backward reactions take place at the same speed, so that macroscopically no visible conversion of substances occurs. On the molecular level, however, reactions continue to take place. Although the macroscopic reaction rate always aims at the chemical equilibrium, equilibrium analysis does not provide information about the time required to reach this state. Information on this is provided by the reaction kinetics.

If one takes into account that in the special case of a chemical equilibrium the reaction proceeds equally fast in both directions, the conversion rate of the concentrations of the individual species becomes zero. Eq. 3.61 results in

$$0 = \alpha \left(k_f \cdot [\mathrm{A}]^\alpha \cdot [\mathrm{B}]^\beta \cdot [\mathrm{C}]^\gamma - k_r \cdot [\mathrm{D}]^\delta \cdot [\mathrm{E}]^\varepsilon \right) \tag{3.63}$$

When transforming the equation as the ratio between forward and backward speed,

$$\rightarrow \frac{k_f}{k_r} = \frac{[\mathrm{D}]^\delta \cdot [\mathrm{E}]^\delta}{[\mathrm{A}]^\alpha \cdot [\mathrm{B}]^\beta \cdot [\mathrm{C}]^\gamma} = K_c \tag{3.64}$$

K_c is a function of the molar concentration of all involved species and is called equilibrium constant.

In order to get an idea of the equilibrium composition of a combustion, an example of octane with the chemical formula C_8H_{18} is given next. According to Burcat, the equilibrium composition of the combustion products and their dependence on the air ratio λ, the temperature and the process pressure is shown (Fig. 3.8).

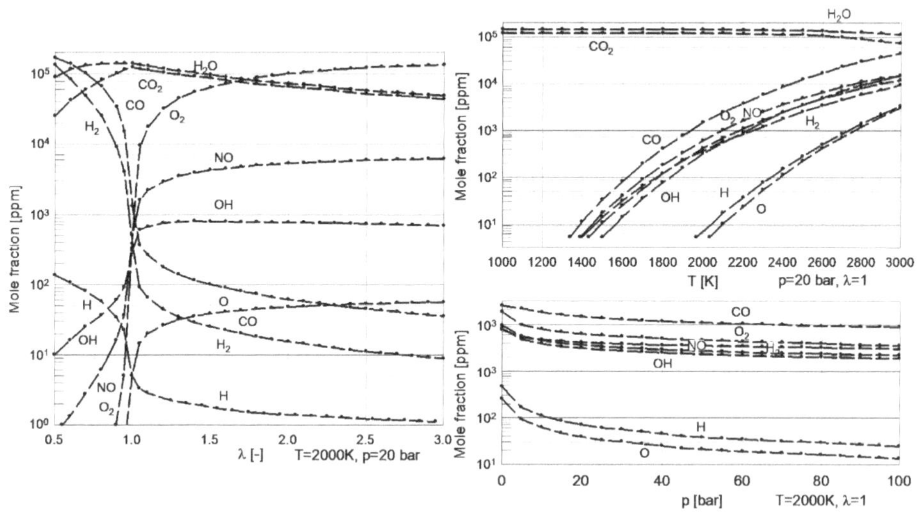

Fig. 3.8 Exhaust gas equilibrium composition for C_8H_{18} fuel according to Burcat. Representation of the dependencies of air/fuel ratio, temperature and pressure

Reaction order

Looking at the reaction equation from 3.65–3.67, this literally means that on the reactant side a number α of A particles collide with β numbers of particles B and γ number of particles C to form D and E on the product side. The probability that the three particles A, B and C collide at the same time and with sufficiently high kinetic energy is very low. It is more likely that two particles first collide, forming an intermediate product and then, if necessary, forming further intermediates to finally form the products D and E. If the overall reaction is broken down into individual steps, their elementary reactions result as follows:

$$\alpha \mathrm{A} + \beta \mathrm{B} \rightarrow \mathrm{A}_\alpha \mathrm{B}_\beta \tag{3.65}$$

$$\mathrm{A}_\alpha \mathrm{B}_\beta + \gamma \mathrm{C} \rightarrow \mathrm{D} \tag{3.66}$$

$$\mathrm{A}_\alpha \mathrm{B}_\beta + \mathrm{D} \rightarrow \mathrm{E} \tag{3.67}$$

Experimentally, it can be determined how the reaction rates of the elementary reactions depend on the respective concentrations of the components A, B, C and D. The dependence of the reaction rate on the exponent, with which the concentration of a certain reactant enters into the law of rates, is called reaction order of the respective reactant. The overall order of a reaction is the sum of the reaction orders of all reactants involved in it. The reaction rate of the individual elementary reactions are multiplied together to give the rate of the total reaction.

Partial balance

In a complex reaction system, it can happen that a large number of reactions take place simultaneously, only some of which proceed so quickly that one can speak of a partial equilibrium of this particular species. The reaction as a whole does not necessarily have to be in chemical equilibrium.

In 1946, Y.B. Zeldovich presented the elementary reaction of nitrogen oxide production for the first time. These describe the formation of thermal nitrogen oxide (NO), which in combustion engines accounts for the largest proportion of all total nitrogen oxides ($\sim$ 90–95% of all NO_x).

$$N_2 + \dot{O} \underset{k_{1,r}}{\overset{k_{1,v}}{\rightleftharpoons}} NO + \dot{N} \tag{3.68}$$

$$\dot{N} + O_2 \underset{k_{2,r}}{\overset{k_{2,v}}{\rightleftharpoons}} NO + \dot{O} \tag{3.69}$$

$$\dot{N} + OH \underset{k_{3,r}}{\overset{k_{3,v}}{\rightleftharpoons}} NO + \dot{H} \tag{3.70}$$

Using a gasoline engine combustion process at stoichiometric condition ($\lambda = 1$), the species concentrations over time are examplarily shown below. After the ignition timing and the corresponding increase of the process temperature $T > 2500$ K, the reactions get activated. The graph shows that the components O and OH are intermediate products that are formed for a short time and then suddenly decomposed again. The partial equilibrium can be represented to determine the species concentration: $dC_O/dt = 0$, $dC_{OH}/dt = 0$ (Fig. 3.9).

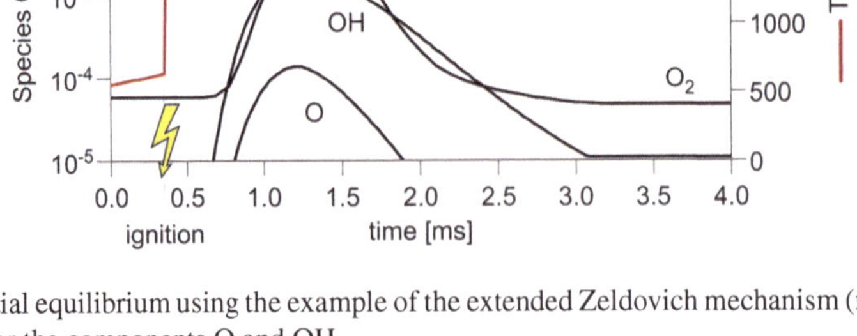

Fig. 3.9 Partial equilibrium using the example of the extended Zeldovich mechanism (nitrogen oxide formation) for the components O and OH

5. Thermodynamics dQ_W: The description of the heat transfer in the cylinder is subject to complex interrelationships. Due to the relatively large proportion and losses that occur on the thermal processes, it is of constant interest to reproduce them with high accuracy on the basis of models in order to identify potential for improvement. The additional difficulty in modeling the thermal process is that a number of consecutive processes as for the charge exchange (fluid dynamic processes), combustion process (chemical processes), internal energy (caloric processes) and kinematics, each introduces an individual error which adds up to an exponential error (looking back to Fig. 3.7). Finally, the summed error has a decisive influence on the calculation of the thermal behavior, so that an accurate prediction becomes a great challenge.

In principle, the heat transfer through the walls Q_w is composed of a convective component Q_e and a radiation component Q_ε:

$$dQ_w = dQ_e + dQ_\varepsilon \tag{3.71}$$

Usually, the radiation part is added to the convective part and is found in the heat transfer coefficient α. The combustion chamber is usually divided at least into three different areas: 1. piston, 2. liner and 3. cylinder head. To achieve a higher quality of resolution, theoretically more sections up to infinitesimal elements can be selected. The valves are usually added to the cylinder head. The description of the heat flow through the cylinder walls follows Newton's equation using the heat transfer coefficient α_i, the transfer area A_i and the temperature difference between the cylinder wall T_w and the gas T_{gas}:

$$dQ_w = \sum_i \alpha_i A_i \left(T_{w,i} - T_{gas,i}\right) \tag{3.72}$$

The determination of a suitable heat transfer coefficient presupposes that the gas temperature in the combustion chamber as well as the wall temperatures are very well represented. For 3D CFD simulations, it is usual to determine temperature distributions of the burned and unburned gas to different cells i ($T_{gas,i}$) over the entire combustion chamber. In 1D, however, the combustion is determined quasi-dimensionally, so that the mean gas temperature is calculated from the local averaging of the gas temperature in the combustion chamber and the index i is omitted (T_{gas}) [5].

For this purpose, usually semi-physical approaches are used, which provide a fast prediction and, at the same time, take into account a sufficiently high degree of boundary conditions. In the literature, there is a large number of presented approaches for the calculation of the heat transfer through the walls [14–17].

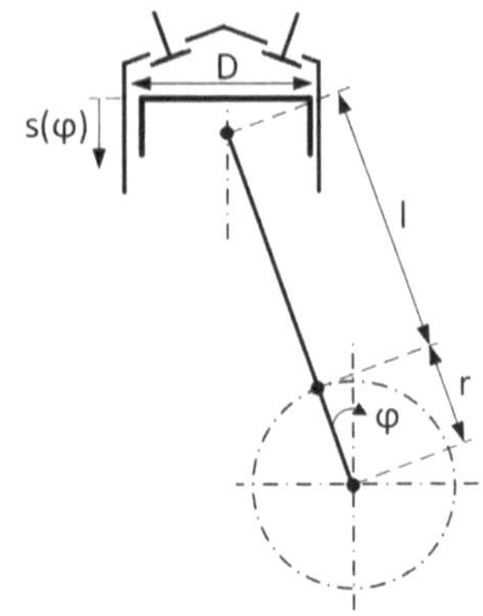

Fig. 3.10 Geometry of the crank mechanism

6. Kinematic processes dW: The mechanism of an engine converts the oscillating motion of the pistons into a rotating motion of the crankshaft (Fig. 3.10). The work dW converted at the piston can be determined through the cylinder pressure and the volume change [5]:

$$dW = -pdV \tag{3.73}$$

Resolving for the crank angle, the work can be derived from the following term. This requires additional inputs as the angular frequency ω, the cylinder bore diameter D and the piston stroke s:

$$\frac{dW}{d\varphi} = -p\omega\frac{dV}{d\varphi} = -p\omega D^2\frac{\pi}{4}\frac{ds}{d\varphi} \tag{3.74}$$

The piston stroke is directly related to the crank angle. The term $\frac{ds}{d\varphi}$ can be calculated as a function of the crank radius r and the conrod ratio $\lambda_s = \frac{r}{l}$:

$$s(\varphi) = r \cdot [[1 - cos\varphi] + \frac{1}{\lambda} \cdot [1 - \sqrt{1 - \lambda_s \cdot sin^2(\varphi)}]] \tag{3.75}$$

$$\frac{ds}{d\varphi} = r\omega \cdot [sin\varphi + \frac{\lambda_s}{2} sin(2\varphi)] \tag{3.76}$$

3.4 Material Strength and Structure

Multi-body simulation

Multi-body simulation (MBS) is a method of numerical simulation, in which simple as well as complex multi-body systems are considered, which are composed of individual components. The composition is realized by simple, rigid or elastic bodies, which are connected by joints, springs, dampers or other contact connections. The MBS is a very useful tool for motion analysis with special focus on the interactions between single components. This enables the determination of loads that result as a consequence of superimposed loads and mass inertia of a system. These loads are used for the early detection of deformations and stresses. New and fast designs of modified kinematics finally help to find optimized solutions. Frequently, MBS tools are used during product development to evaluate the characteristics of safety, comfort, durability and performance.

Basically, the MBS is divided into kinematic and dynamic calculations. In a kinematic view, the systems have no dynamic degrees of freedom. Since the change over time is insignificant here, stationary observations are sufficient to evaluate these. For more complex problems, on the other hand, it is often not sufficient to regard bodies as rigid in order to meet higher accuracy requirements. In this respect, the structural elasticity of the components is included, so that additional, deformation-induced stresses are taken into account. A more detailed representation of a rigid MBS can be realized by integrating additional simulation methods such as the finite element method (FEM), flow simulation (CFD) and thermodynamics. Along with minor additional modeling effort and corresponding higher computational effort, the predictions are much more accurate over deformations, dynamics and stresses of components.

Finite Elemente

Finite Element Analysis (FEM) is a numerical calculation method for the investigation of the strength and deformation of bodies with static and dynamic load conditions. FEM has become widely accepted in the development of mechanical components for industrial applications over the last decades and is indispensable today. In the application of Finite Element Analysis (FEA) software, CAD-manufactured components are imported and divided into a multitude of small finite units (mesh). If the investigation of a load on critical zones of a body is of high interest, these selected areas are resolved higher than others, so that detailed results can be obtained. Depending on the choice of the resolution (shape function) of the smallest volume elements (tetrahedron, hexahedron, octahedron, etc.), deeper insights into the structural behavior can be obtained at the expense of more computational time.

The application possibilities of FEM have expanded massively, especially in the last few years, thanks to the availability of high-performance computers. This allows to increasingly investigate coupled problems, such as the combination with multi-body systems and the interaction between structures and fluids from (CFD), acoustics, thermomechanics, thermochemistry, ferroelectrics, electromagnetics and other relevant areas. FEM applications are basically divided into static, dynamic and modal problems. With the help of static analyses, linear static and nonlinear quasi-static structures can be analyzed. With a dynamic analysis, however, oscillating or unsteady loads of components can be analyzed with respect to their structure. Modal analysis is used when the focus is on determining the eigenfrequencies of a structure through vibration excitation and the prevention of undesired interference.

Fig. 3.11 2D and 3D finite elements with linear, quadratic and cubic deformation

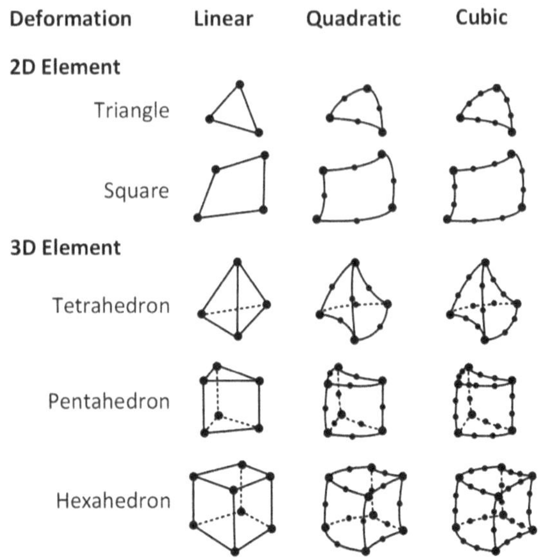

In order to examine bodies of any kind with corresponding materials and arbitrary fields of application for load, deformation and durability, they are divided into finite elements for an FEM investigation. This results in a so-called "mesh" of the body. The finer the mesh is divided into finite (finitely small) elements, the higher the resolution of the results, which is directly related to a higher calculation effort. Due to their effectiveness, triangular and quadrangular elements have become widespread for conventional software for 2D surfaces. For 3D bodies, on the other hand, tetrahedral, pentahedral and hexahedral elements as shown in Fig. 3.11 are used. If the deformation of bodies is to be calculated in addition to the load, quadratic or cubic volumes can be selected.

Figure 3.12 shows on a 2D connection beam what a possible transformation of a body into finite elements can look like. The element is fixed to the left and the bottom side. A meshing is exemplarily done with triangular elements (quadratic) with a coarse mesh (center) and a finer mesh (right).

Fig. 3.12 Finite element for a 2D surface with a large and a small mesh size

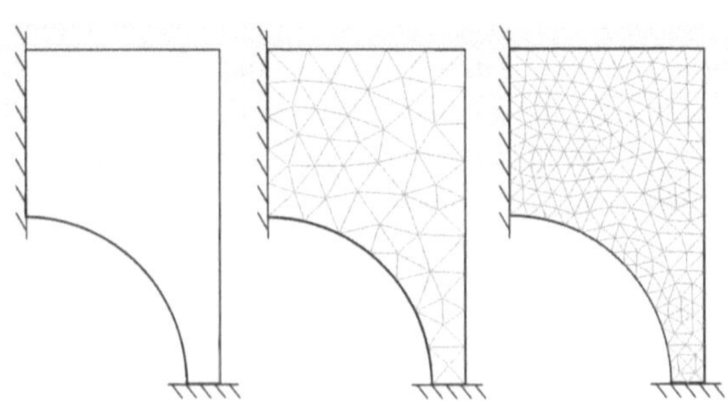

FEM mainly deals with the solution of boundary value problems. Boundary value problems are problems for which solutions are sought for a given differential equation so that certain conditions arise at the boundaries of the corresponding defined range. Basically, two different methods can be chosen, the application of a force balance or the that of an energy balance. In conventional FEM simulation tools, both methods are provided. Hence, the plausibility of the boundary conditions of an approach selected at the beginning can be checked afterwards by means of an energy balance approach.

In principle, FEM methods can be used to analyze mechanical stress, thermal stress or examine electrical and magnetic flux over solids. The process of an FEM analysis always follows one scheme:

- Definition of a boundary value problem (2D, 3D boundary conditions);
- Creating a model (importing a CAD file);
- Definition of all loads (static or dynamic load);
- Meshing of the model (resolution of the cells, linear, quadratic or cubic deformation);
- Definition of the analysis method (force balance, energy balance).

After a component has been divided into small finite elements, the balances of forces are established at each element and the equations of elasticity theory are applied.

3.4.1 Theory of Elasticity

Using a section of a cubic volume, which in its deformed state corresponds to a hexahedron, all relevant forces and stresses are shown. The normal stresses, which are perpendicular to the squared surfaces, are called $\sigma_x, \sigma_y, \sigma_z$. The shear stresses pointing in the surface direction are called $\tau_{xy}, \tau_{yz}, \tau_{xz}$. Together with the acting forces f_x, f_y, f_z on the volume, the system can be defined and set up as a boundary value problem.

As a result of the elasticity of a body, the forces acting on a volume element lead to deformation and thus to a displacement of the geometry. The determination of the displacement of the individual volumes and hence of the sum of all volume elements forming a body is the actual focus of each FEA calculation. Four steps of the theory of elasticity can be summarized to determine deformations:

1. The force and momentum balance;
2. The strain-displacement relation $\varepsilon - u$ and the material constancy;
3. The stress-strain relation $\sigma - \varepsilon$;
4. Boundary value problem and displacement function u(x);
5. Determining the stiffness matrix [k] and displacement vector {d}.

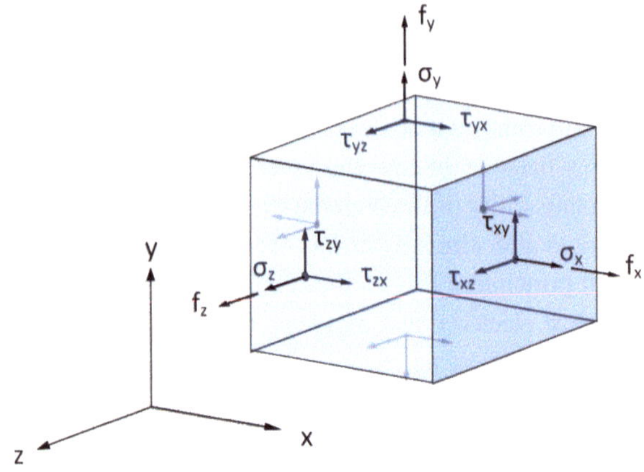

Fig. 3.13 Force balance on a body

1. Force and momentum balance

If one sets up the equations for the equilibrium of forces and moments according to Fig. 3.13, a total of 6 differential equations and 9 unknown quantities (τ_{xy}, $\tau_{yz}\tau_{xz}$, $\tau_{zx}\tau_{yx}$, τ_{zy}, σ_x, σ_y, σ_z) are obtained. The boundary value problem assumes that the forces f_x, f_y, f_z acting on the volume are known. Balance of power:

$$0 = f_x + \frac{\partial \sigma_x}{\partial x} + \frac{\partial \tau_{yx}}{\partial y} + \frac{\partial \tau_{zx}}{\partial z} \quad (1)$$

$$0 = f_y + \frac{\partial \sigma_y}{\partial y} + \frac{\partial \tau_{xy}}{\partial x} + \frac{\partial \tau_{zy}}{\partial z} \quad (2)$$

$$0 = f_z + \frac{\partial \sigma_z}{\partial z} + \frac{\partial \tau_{xz}}{\partial x} + \frac{\partial \tau_{yz}}{\partial y} \quad (3)$$

Equilibrium of moments:

$$\tau_{xy} = \tau_{yx} \quad (4)$$

$$\tau_{yz} = \tau_{zy} \quad (5)$$

$$\tau_{xz} = \tau_{zx} \quad (6)$$

Equations (4), (5) and (6) ensure that 3 of the stress variables are eliminated, leaving 3 differential equations (1), (2), (3) and the 6 unknown variables (τ_{xy}, τ_{yz}, τ_{xz}, σ_x, σ_y, σ_z).

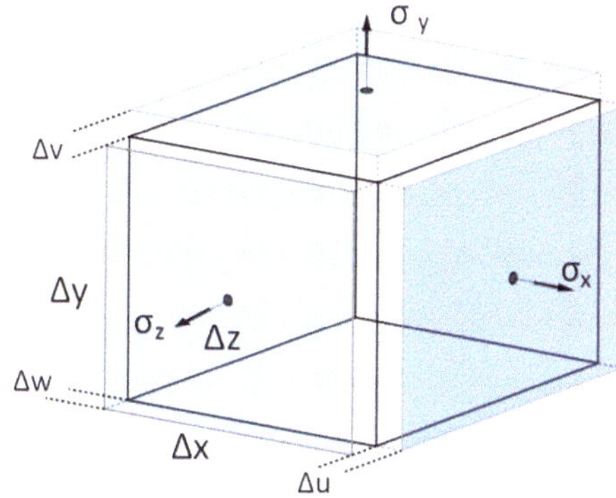

Fig. 3.14 Deformation of a body

2. The strain-displacement relation $\varepsilon - u$ and material constancy

If forces act on a body, a deformation occurs in the direction of the applied force. The deformation is approximately defined as the ratio between the deformation length δu and the total length δx, as shown in Fig. 3.14. Looking at an infinitesimal volume element, the strain can be represented as a partial derivative. This results in 3 new differential equations with 6 unknown quantities, namely the 3 strain quantities (ε_x, ε_y, ε_z) and the 3 deformation variables (u, v, w).

$$\varepsilon_x = \frac{\partial u}{\partial x} \quad (7)$$

$$\varepsilon_y = \frac{\partial v}{\partial y} \quad (8)$$

$$\varepsilon_z = \frac{\partial w}{\partial z} \quad (9)$$

In addition to the deformation, there is also an angular deformation of the volume element as a result of the acting forces, if the volume is not fixed at the edges. In Fig. 3.15, a volume element is fixed at the lower left edge (A-B). If the normal stress σ_x, σ_y, σ_z act on the body, the body deforms with the angle φ_1 due to σ_y, with φ_2 due to σ_x and φ_3 due to σ_z. From this, 3 differential equations for the shear deformations γ_{xy}, γ_{yz}, γ_{xz} can be set up, which result from the forces in the plane.

$$\gamma_{xy} = \varphi_1 + \varphi_2 + \frac{\partial v}{\partial x} + \frac{\partial u}{\partial y} \quad (10)$$

$$\gamma_{yz} = \varphi_2 + \varphi_3 + \frac{\partial u}{\partial y} + \frac{\partial w}{\partial y} \quad (11)$$

$$\gamma_{xz} = \varphi_1 + \varphi_3 + \frac{\partial v}{\partial x} + \frac{\partial w}{\partial y} \quad (12)$$

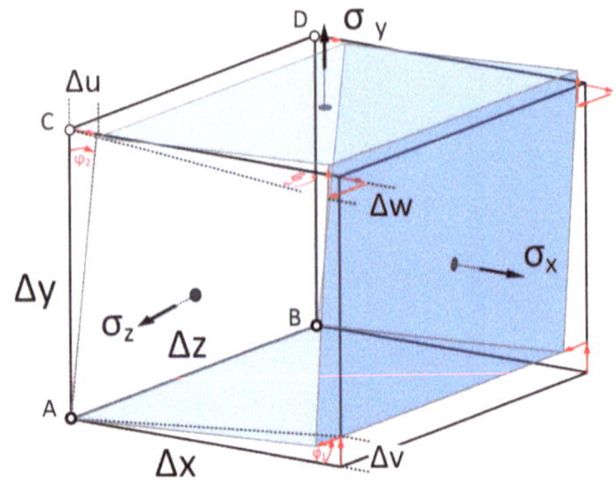

Fig. 3.15 Shear deformation of a body

For practical reasons, the strain variables and shear stresses are combined in a strain vector as shown in Eq. 3.77. The partial derivatives of the deformations are separated from each other on the right side of the equation and divided into a partial derivative operator matrix and deformation vector.

$$\underbrace{\begin{pmatrix} \varepsilon_x \\ \varepsilon_y \\ \varepsilon_z \\ \gamma_{xy} \\ \gamma_{yz} \\ \gamma_{xz} \end{pmatrix}}_{strain\ vector} = \underbrace{\begin{pmatrix} \frac{\partial}{\partial x} & 0 & 0 \\ 0 & \frac{\partial}{\partial y} & 0 \\ 0 & 0 & \frac{\partial}{\partial z} \\ \frac{\partial}{\partial y} & \frac{\partial}{\partial x} & 0 \\ 0 & \frac{\partial}{\partial z} & \frac{\partial}{\partial y} \\ \frac{\partial}{\partial z} & 0 & \frac{\partial}{\partial x} \end{pmatrix}}_{operator\ matrix} \cdot \underbrace{\begin{pmatrix} u \\ v \\ w \end{pmatrix}}_{deformation\ vektor} \tag{3.77}$$

The relationship between strain and displacement can be summarized in the vectorial notation $\varepsilon = [\partial] \cdot u$. In total, this relationship results in 9 unknown quantities ($\varepsilon_x, \varepsilon_y, \varepsilon_z, \gamma_{xy}, \gamma_{yz}, \gamma_{xz}, u, v, w$) and 6 differential equations.

3. The stress-strain relation $\sigma - \varepsilon$

The stress-strain behavior is represented by Hooke's law, which describes a linear relationship between the stress σ and the strain ε by the modulus of elasticity E. By definition, the law applies to materials only in the linear elastic range. As soon as plastic deformation occurs, the physical phenomena behind this follow much more complex laws:

$$\sigma = E \cdot \varepsilon \tag{3.78}$$

Another important material constant comes into play through the Poisson number ν, which is a measure of lateral contraction. The Poisson effect follows the law of material continuity, as it describes the tendency of a material to expand perpendicular to the compression direction. On the other hand, when a material is stretched, it tends to contract transversely to the stretched direction.

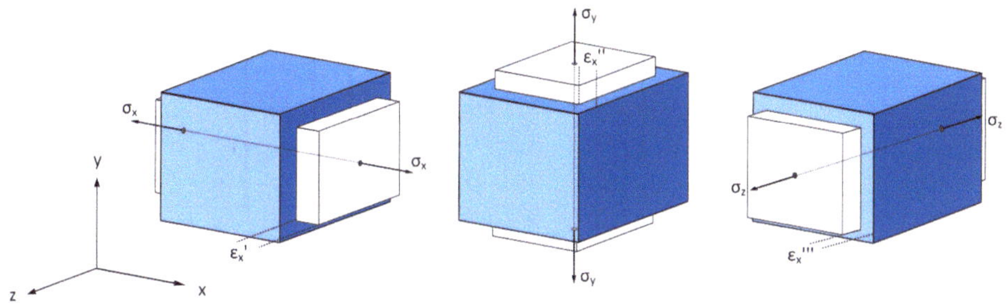

Fig. 3.16 Transversal contraction of a body

Figure 3.79 shows how a body expands in the linear-elastic range under consideration of the transverse contractions when it is exposed to a normal stress. When force is applied in the x-direction, the body expands by the strain ratio $\varepsilon_x' = \frac{\sigma_x}{E}$, when force is applied in the y-direction, it contracts by $\varepsilon_x'' = -\nu\frac{\sigma_y}{E}$ and when force is applied in the z-direction, it contracts by $\varepsilon_x''' = -\nu\frac{\sigma_z}{E}$. The total contraction of a body in the x-direction is composed of the individual strains (Fig. 3.16):

$$\varepsilon_x = \varepsilon_x' + \varepsilon_x'' + \varepsilon_x''' \tag{3.79}$$

If one proceeds equally for the coordinates y and z, the transverse contraction can be described as follows:

$$\varepsilon_x = \frac{1}{E}\left[\sigma_x - \nu\left[\sigma_y + \sigma_z\right]\right] \quad (13)$$

$$\varepsilon_y = \frac{1}{E}\left[\sigma_y - \nu\left[\sigma_x + \sigma_z\right]\right] \quad (14)$$

$$\varepsilon_z = \frac{1}{E}\left[\sigma_z - \nu\left[\sigma_x + \sigma_y\right]\right] \quad (15)$$

Similarly, the shear deformations γ can be related to the shear stresses τ, which results in 3 further equations:

$$\gamma_{xy} = \frac{\tau_{xy} \cdot 2\,(1+\nu)}{E} \quad (16)$$

$$\gamma_{yz} = \frac{\tau_{yz} \cdot 2\,(1+\nu)}{E} \quad (17)$$

$$\gamma_{xz} = \frac{\tau_{xz} \cdot 2\,(1+\nu)}{E} \quad (18)$$

For practical reasons, the vectorial notation is often used here as well:

$$\begin{pmatrix} \sigma_x \\ \sigma_y \\ \sigma_z \\ \tau_{xy} \\ \tau_{yz} \\ \tau_{xz} \end{pmatrix} = \begin{pmatrix} \frac{1-\nu}{1-2\nu} & \frac{\nu}{1-2\nu} & \frac{\nu}{1-2\nu} & 0 & 0 & 0 \\ 0 & \frac{1-\nu}{1-2\nu} & 0 & 0 & 0 & 0 \\ 0 & 0 & \frac{1-\nu}{1-2\nu} & 0 & 0 & 0 \\ 0 & 0 & 0 & \frac{1}{2} & 0 & 0 \\ 0 & 0 & 0 & 0 & \frac{1}{2} & 0 \\ 0 & 0 & 0 & 0 & 0 & \frac{1}{2} \end{pmatrix} \cdot \begin{pmatrix} \varepsilon_x \\ \varepsilon_y \\ \varepsilon_z \\ \gamma_{xy} \\ \gamma_{yz} \\ \gamma_{xz} \end{pmatrix} \tag{3.80}$$

The stress-strain ratio can be represented in the vectorial notation: $\{\sigma\} = [D]\{\varepsilon\}$. The matrix $[D]$ is a characteristical material matrix or also called the elasticity matrix.

By the laws of stress-strain relations, 6 further equations without additional unknowns are created. Together with the equilibrium of forces and momentums from (1.) and the ratio between strain and deformation from (2.), a total of 18 equations with 18 unknown variables are combined, so that a solvable system of equations results and finally the displacement vector $\{d\}$ can be solved.

4. Boundary value problem and displacement function *u(x)*

To solve the differential equation, an assumption about the displacement field has to be made. Typically, linear displacement fields are assumed for beam elements with a function $u(x) = a_0 + a_1 \cdot x$, see Fig. 3.17. In a mesh, common FEA software offer several possibilities for selection. Using a hexahedron as an example, it is shown that a linear approach, a quadratic, a cubic or a function of a higher order is possible in principle. However, when the order increases, the operations and thus the calculation time increase exponentially as a resulting effect.

5. Determination of the stiffness matrix [*k*] and displacement vector {*d*}

After having defined boundary values for the deformation of all cells in (4.), the last step is to determine the displacement of the nodes for a finite element and, in summary, the displacement of all nodes of an overall system consisting of many finite elements.

If we consider a 2D finite element with a rectangular profile and a linear displacement function (see Fig. 3.17), which is loaded with the forces F_x and F_y, all nodes are shifted by the displacement vector $\{d_i\}$.

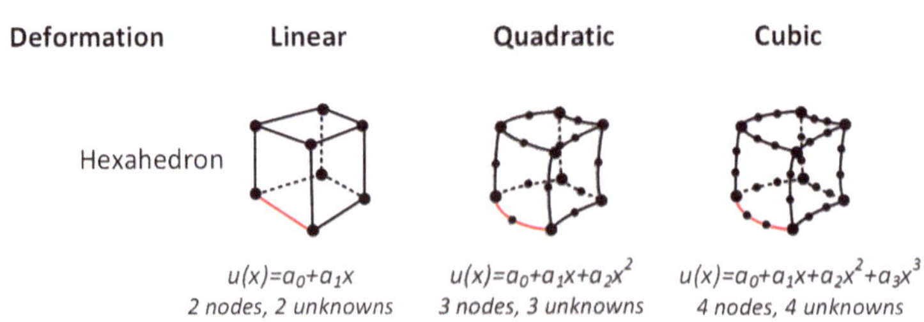

Fig. 3.17 Number of nodes per body edge by a displacement function

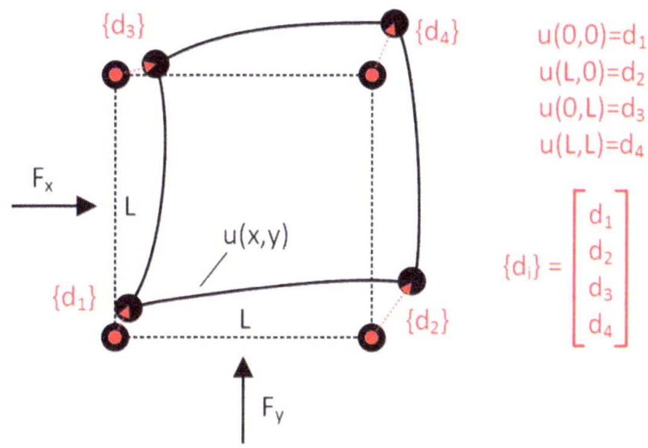

Fig. 3.18 Displacement of all nodes of a body represented by the displacement vector $\{d_i\}$

Together with the stiffness matrix $[k]$, the system of linear equations represents a simplified relationship between loads and displacements in each element for a node i, which must be solved to obtain an approximate solution of the differential equations (1)–(18):

$$\{f\}_i = [k]_i \cdot \{d\}_i \tag{3.81}$$

Finally, the law of continuity is applied to the entire system. This ensures that loads and displacements are connected between all elements and that there is no material gap or overlap within a system:

$$\{F\} = [K] \cdot \{D\} \tag{3.82}$$

The solution of the nodal displacement is found by inverting the matrix solving the equation for $\{D\}$:

$$\{D\} = [K]^{-1} \cdot \{F\} \tag{3.83}$$

Once the displacement vector $\{D\}$ has been determined, all stresses (normal stresses and shear stresses) as well as strains and shear deformations can be determined in each element node by means of the equations presented above [18].

3.4.2 Alternative Methods

In addition to the presented method of solving a boundary value problem by means of the balance of forces, alternative possibilities are also available. The most popular ones are the energy method (Minimal Potential Theory), the principle of virtual work or the Rayleigh-Ritz approximation. In the following, the example of the minimal potential theory is used to show how an identical problem can be solved by applying an energy balance [19].

Minimal Potential Theory

If a body is applied with one or more loads, internal tensions arise in the body. In its entirety, the potential energy π_p of the body can be described as the sum of the internal energy U and the potential of internal and external forces Ω:

$$\pi_p = U + \Omega \tag{3.84}$$

The deformation of a body corresponds to the internal energy and, according to the law of conservation of energy, this must be in balance with the external energy. In this respect, the system of equations can be solved by an extreme value problem.

Of all possible deformations of a body, the equilibrium deformation has the lowest total potential energy, or in other words: The total potential energy of a system is stationary when the system is in equilibrium and the change in potential energy becomes minimal ($\delta\pi_p = 0$). Since the total potential energy depends on the individual deformations of the body, i.e. on the deformations of all nodes, it can be described as a function of the degrees of freedom (node displacements):

$$\pi_p = \{d_1, d_2 \dots d_n\} \tag{3.85}$$

Using the chain rule for the derivation of nested functions, the derivation of π_p results in

$$\delta\pi_p = \frac{\partial\pi_p}{\partial d_1}\delta d_1 + \frac{\partial\pi_p}{\partial d_2}\delta d_2 + \dots + \frac{\partial\pi_p}{\partial d_n}\delta d_n \tag{3.86}$$

This means that for the condition delta $\delta\pi_p = 0$, all individual partial derivations must be zero:

$$\frac{\partial\pi_p}{\partial d_1} = 0, \frac{\partial\pi_p}{\partial d_2} = 0, \dots, \frac{\partial\pi_p}{\partial d_n} = 0 \tag{3.87}$$

The following presentation summarizes the conditions:

$$\frac{\partial\pi_p}{\partial\{d\}} = 0 \tag{3.88}$$

This forms a system of n equations, with which the displacement of the nodes must be determined. The internal energy U of a stressed body is calculated from the integral of the internal stress and the strain of the volume:

$$U = \frac{1}{2}\int_V \{\sigma\}^T\{\varepsilon\}dV \tag{3.89}$$

Compared to Eq. 3.78, an energetic consideration of the stress σ requires an additional stress σ_0, which describes the energetic condition at time 0.

$$\{\sigma\} = \{\sigma_0\} + [D](\{\varepsilon\} - \{\varepsilon_0\}) \tag{3.90}$$

According to the product rule for matrices, the transposed equation is obtained after dissolving the bracket:

$$\{\sigma\}^T = \underbrace{\{\sigma_0\}^T}_{\text{initial stresses}} + \underbrace{\{\varepsilon^T\}[D]^T}_{\text{strains from loads}} - \underbrace{\{\varepsilon_0\}^T[D]^T}_{\text{thermal strains}} \tag{3.91}$$

Using Eq. 3.89, the inner energy U results in

$$U = \underbrace{\frac{1}{2}\int_V \{\sigma_0\}^T\{\varepsilon\}dV}_{\text{initial stresses}} + \underbrace{\frac{1}{2}\int_V \{\varepsilon\}^T[D]^T\{\varepsilon\}dV}_{\text{strains from loads}} - \underbrace{\frac{1}{2}\int_V \{\varepsilon_0\}^T[D]^T\{\varepsilon\}dV}_{\text{thermal strains}} \tag{3.92}$$

If we draw up the energy balance for the potential of the forces Ω, which is applied on the system from outside, 3 terms have to be considered—1. forces acting on the body, 2. surface tractions and 3. loads acting on the nodes:

$$\Omega = \underbrace{-\int_V \{u\}^T\{f_B\}dV}_{bodyforce} - \underbrace{\int_S \{u_S\}^T\{f_S\}dS}_{\text{surface tractions}} - \underbrace{\{d\}^T\{f_P\}}_{\text{nodal point loads}} \tag{3.93}$$

In summary, the total potential energy is obtained by inserting the terms U and Ω in Eq. 3.84:

$$\pi_P = \underbrace{\frac{1}{2}\int_V \{\sigma_0\}^T\{\varepsilon\}dV}_{\text{initial stresses}} + \underbrace{\frac{1}{2}\int_V \{\varepsilon\}^T[D]\{\varepsilon\}dV}_{\text{strains from loads}} - \underbrace{\frac{1}{2}\int_V \{\varepsilon_0\}^T[D]\{\varepsilon\}dV}_{\text{thermal strains}}$$

$$- \underbrace{\int_V \{u\}^T\{f_B\}dV}_{\text{body force}} - \underbrace{\int_S \{u_S\}^T\{f_S\}dS}_{\text{surface tractions}} - \underbrace{\{d\}^T\{f_P\}}_{\text{nodal point loads}} \tag{3.94}$$

According to Fig. 3.18, the deformation $\{u\}$ can be projected by the node matrix $[N]$ onto the displacement vector $\{d\}$ of the nodes by the equation $\{u\} = [N]\{d\}$.

$$\{\varepsilon\} = [\partial]\{u\} = [\partial][N]\{d\} = [B]\{d\} \tag{3.95}$$

By introducing the partial derivative of the nodal matrix $[\partial][N] = [B]$, the following simplified notation for the potential energy results:

$$\pi_P = \frac{1}{2}\int_V \{\sigma_0\}^T[B]\{d\}dV + \frac{1}{2}\int_V \{d\}^T[B]^T[D][B]\{d\}dV - \frac{1}{2}\int_V \{\varepsilon_0\}^T[D][B]\{d\}dV$$

$$- \int_V \{d\}^T[N]^T\{f_B\}dV - \int_S \{d\}^T[N_S]^T\{f_S\}dS - \{d\}^T\{f_P\} \tag{3.96}$$

If one excludes the vector $\{d\}$ from the integral and applies the equation $\frac{\partial \pi_P}{\partial\{d\}} = \{0\}$ from 3.88 to determine the extremum problem, the result is

$$\{0\} = \left(\frac{1}{2}\int_V \{\sigma_0\}^T\,[B]dV\right)\{d\} + \frac{1}{2}\{d\}^T\left(\int_V [B]^T[D][B]dV\right)\{d\} - \left(\frac{1}{2}\int_V \{\varepsilon_0\}[D][B]dV\right)\{d\}$$
$$-\{d\}^T\left(\int_V [N]^T\{f_B\}dV\right) - \{d\}^T\left(\int_S [N_S]^T\{f_S\}dS\right) - \{d\}^T\{f_P\} \quad (3.97)$$

When applying some basic rules of linear algebra for the multiplication of vectors and matrices and summarizing the displacement vector $\{d\}$, the following final form of the equation is obtained, which now has the same characteristics as the equation for the stiffness matrix from 3.81 from the elasticity theorem. Finally, the displacement vector for a finite element or the displacement vector $\{d\}$ for a body in the sum of many finite elements can be determined as

$$\underbrace{\frac{1}{2}\int_V \{\sigma_0\}^T\,[B]dV - \frac{1}{2}\int_V \{\varepsilon_0\}[D][B]dV - \int_V [N]^T\{f_B\}dV - \int_S [N_S]^T\{f_S\}dS - \{f_P\}}_{\{f\}}$$
$$= \underbrace{\left(\int_V [B]^T[D][B]dV\right)}_{[k]} \cdot \underbrace{\{d\}}_{\{d\}} \quad (3.98)$$

3.5 Acoustics

In recent years, numerical acoustics has developed rapidly in parallel with the increasing performance of computers and permeates almost all fields of acoustics. Among the most widespread wave-theoretical methods of numerical acoustics are the boundary element method, the finite element method and the beam tracing method.

Generally, a distinction is made between problems of acoustics in gases and solid-state sound. It is obvious that the acoustics of gases are based on the fundamentals of fluid dynamics. In contrast, the theory of solid-state sound refers to the calculation principles that are also used for material structures [20].

A common method used in acoustics is the Fourier analysis. This method is mainly applied to perform spectral analyses of a sound wave, especially to generate or reproduce synthetic signals.

According to the Fourier theorem, a periodic signal $f(t)$ with the period T can be represented by a constant component and an infinite sum of harmonic signals $h_i(t)$ with different angular frequencies ω_i, which differ in their amplitudes c_i and phases φ_i. The angular frequencies of the suborders form multiples of the basic angular frequency $\omega_0 = \frac{2\pi}{T}$. For special periodic functions, the sum representation, also called trigonometric series or Fourier series, can be represented as follows:

$$f(t) = c_0 + \sum_{i=1}^{\infty} [a_i \cos(i\omega_0 \cdot t) + b_i \sin(i\omega_0 \cdot t)] \quad (3.99)$$

The quantities c_0, a_i and b_i are called Fourier coefficients. c_0 represents the mean value (direct component) of the signal $f(t)$. If $f(t)$ stands for a pressure $p(t)$ oscillating in time, c_0 corresponds to the mean value of the pressure signal. The representation of the Fourier series can be simplified if the following relationship is used:

$$a_i \cos(i\omega_0 \cdot t) + b_i \sin(i\omega_0 \cdot t) = c_i \cos(i\omega_0 \cdot t + \varphi_i) \quad (3.100)$$

$$c_i = \sqrt{a_i^2 + b_i^2} \quad (3.101)$$

$$\varphi_i = arctan\left(\frac{a_i}{b_i}\right) \quad (3.102)$$

Herewith, Eq. 3.99 becomes the so-called spectral representation of the Fourier series:

$$f(t) = c_0 + \sum_{i=1}^{\infty} c_i \cos(i\omega_0 \cdot t + \varphi_i) \quad (3.103)$$

A periodic signal $f(t)$ can thus be represented by a Fourier analysis with the following quantities [21]:

c_0 : Constant component (mean value of the signal $f(t)$);
c_i : Amplitude of order i;
φ_i: Phase of order i.

For a periodic signal, the transformation from the time domain into its spectral domain uses both an integral and a complex notation:

$$F(\omega) = \int_{-\frac{T}{2}}^{\frac{T}{2}} f(t)e^{-i\omega t}\mathrm{d}t \quad (3.104)$$

The back transformation from the spectral to the time domain is

$$f(t) = \int_{-\frac{T}{2}}^{\frac{T}{2}} f(\omega)e^{+i\omega t}\mathrm{d}\omega \quad (3.105)$$

The function $f(t)$ to be transformed is often not known, but can only be tapped at N discrete times with $t_k = (N-1) \cdot \Delta t$. In such cases the discrete Fourier transformation (DFT) is used. DFT assumes that $f(t)$ uses a periodical continuation outside the interval. The Fourier coefficients are calculated according to the definition as follows:

$$F(n) = \frac{1}{N} \sum_{n=0}^{N-1} f_n e^{\frac{-2\pi i}{N}} \tag{3.106}$$

The inverse discrete Fourier transform (IDFT) is used for the reverse transformation from the spectral to the time domain is

$$f(n) = \frac{1}{N} \sum_{n=0}^{N-1} F_n e^{\frac{+2\pi i}{N}} \tag{3.107}$$

Based on Cooley et al., an algorithm was developed to reduce the number of complex arithmetic operations to calculate the spectral lines of the DFT by the factor $\frac{N}{\ln N}$ [22]. Due to the lower computational effort and the resulting shorter calculation process, the numerically favorable execution rule of the DFT is called Fast Fourier Transformation (FFT) in this context [23].

Figure 3.19 shows signals of different amplitudes and orders of a fundamental frequency f_{eigen} (here the orders 1, 3, 5, 7) and projects them on the time domain, where the superposition of all single orders can be recognized. At the same time, a projection onto the frequency domain takes place, which provides a breakdown of all involved signals.

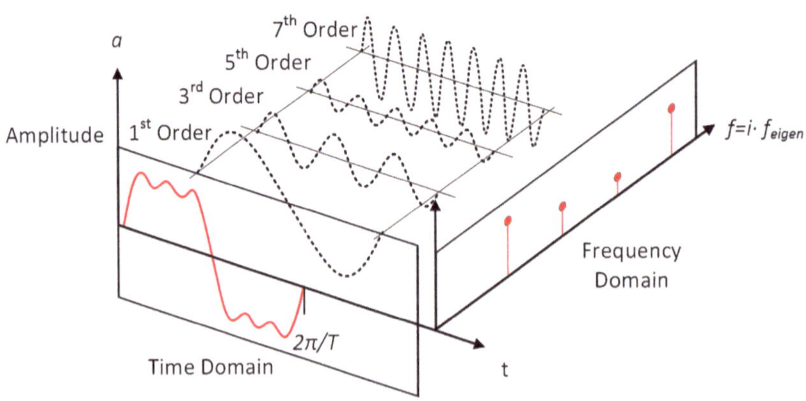

Fig. 3.19 Time and frequency spectrum by superposition of harmonic signals

Big Data for Powertrains

4

4.1 The Future of Powertrain Concepts

Tighter CO_2 limits as a result of climate targets, the Germany-wide diesel emission scandal, Brexit and, last but not the least, the trade conflict between the USA and China have paralyzed the automotive industry worldwide. All the signs are suddenly pointing to a new era and a colossal reshuffle. Manufacturers, suppliers and service providers are asking themselves what is the right strategy to get out of the situation as smoothly as possible.

Basically, the automotive industry reflects the transformation turn of digitization in all its far-reaching facets. As a pioneer of digitization and all the opportunities that this offers, the U.S. has taken on a brilliant role and demonstrated to the world the power behind modern corporate development and lean management style. Horizontal value creation chains are getting ever more popular and inexorably permeating today's corporate structures. Not only for modern companies but even more so for conservative ones it has become inevitable to overhaul old structures and adapt highly effective ones in order to remain viable in the highly competitive market. It is now the turn of the automotive industry to be put to the test, because it currently seems as if the cards are being reshuffled. The right positioning on the market today can prove to be very successful in a few years and a wrong positioning will lead companies to even greater problems.

Which positioning is the right one?

Digitization begins primarily in the mind. This means a change of thought patterns and, as a consequence, a change of an entire society. Ever-increasing personalization has led to a greater characterization of individuals. In this context, the term "granularity" of humanity is often used. Opinions can be expressed more easily than ever through social media. Collectively, they are able to influence and manipulate the opinions of others, even overshadowing expert opinions despite the lack of knowledge, thanks to media forces. As a result, never

A. Mirfendreski, *Powertrain Development with Artificial Intelligence*,
https://doi.org/10.1007/978-3-662-63863-7_4

before has there been such a large discrepancy of opinions among experts, society and, at the same time, politics.

Change means that there will be winners, just as there must be losers. For the sake of debate, let's say the combustion engine seems to have become obsolete because its potential is coming to an end and is damaging the climate. However, there are no short-term alternatives available, only medium-term visions. Is it possible to produce a high proportion of electricity from renewable sources worldwide to justify electric mobility through green driving? Are there sufficient lithium and cobalt deposits to cover the global battery demand? And can these raw materials be mined and sustainably disposed of in a resource-oriented manner and under fair working conditions? Doesn't it make more sense to leave the entire transportation infrastructure as it is and switch to synthetic and biologically produced fuels in order to continue to operate our valuable combustion engines? These are questions that every car manufacturer must answer for itself in order to set the strategic course for its future.

Where are we heading?
The world sales for automobiles had reached an interim peak of 85 million in 2018. The EU commission has passed very challenging CO_2 limits by 2030, with great consequences for all car manufacturers and their employees. Compared to the current level of 118 gCO_2/km, newly registered vehicles may only emit 95 gCO_2/km from 2021 and 59 gCO_2/km from 2030 onwards. In addition to the agreed CO_2 values of the EU Commission, Fig. 4.1 shows the CO_2 emission limits of other countries, which were passed in June 2019 and published by the International Council on Clean Transportation (ICCT) [24].

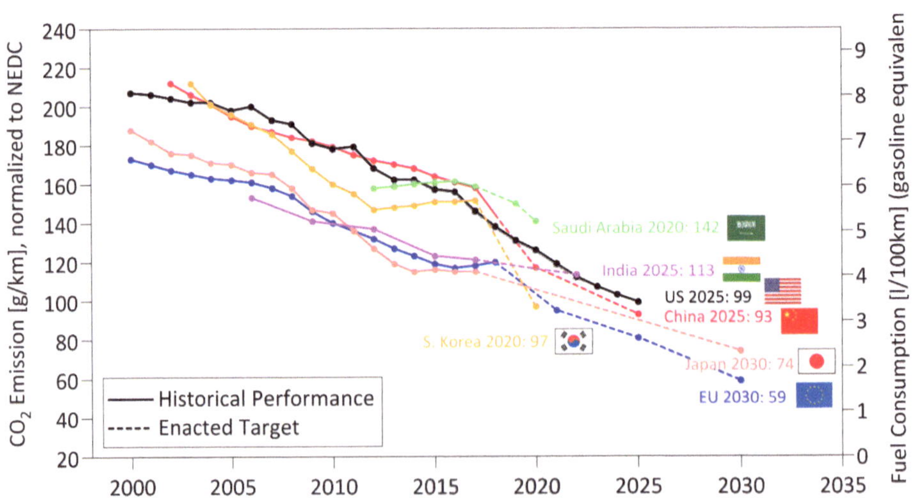

Fig. 4.1 Emission limits according to the European exhaust standard [25]

Despite the ambitious climate protection laws and efficiency targets set by the German government, mobility will stay important for all social classes and remain an expression of personal freedom. Climate protection will inevitably entail higher costs for the population, so it is important to carefully examine and weigh up the advantages and disadvantages of all types of powertrains and energy sources. Today, a large number of renowned studies provide possible scenarios of tomorrow's mobility, which serve as important guidelines supporting understanding which CO_2 targets can realistically be achieved. Which driving concepts can make a significant contribution to this and how do political decisions affect the new car fleet for vehicle customers?

All presented scenarios show that even assuming a value of 95 gCO_2/km as the EU limit in 2020, it is possible to halve the CO_2 emissions of the car fleet down to 45 gCO_2/km without switching to new forms of transport. High-quality fuels (including synthetic fuels and biofuels) in combination with innovative combustion engines enable efficiency improvements in the fleet, both by conventional powertrain systems and by the wide range of hybrid variants. The increasing proportion of electric driving, the growing shares of alternative powertrain systems such as the fuel cell, should help to achieve the defined targets. A major requirement for a growing share of electric mobility on the market is the provision of low-CO_2 electricity, i.e. an electricity mix with a steadily increasing portion of regeneratively generated energy.

The renowned studies selected in the following are intended to serve as guides showing where the automotive landscape is heading in the future: 1. "The Passenger Car Market until 2040" by the German Aerospace Center (DLR), 2013, a study from the perspective of a mineral oil company; 2. "Passenger Car Scenarios until 2040" by Shell from 2014; and 3. "Climate Protection Contribution of Transport until 2050" by the German Federal Environment Agency (UBA), 2016.

New registrations

For new registrations, DLR proposes a moderate "trend"-scenario with a target value of 70 gCO_2/km by 2040, in which conventional engines will hold higher market shares in the long term. The assumptions made here are stable environmental conditions, moderate change in consumer preferences, steady technological progress, a gradually changing fuel mix and a further decreasing emissions target path. Access to automobility in terms of affordability remains an important criterion for this scenario.

In contrast to this, there is an ambitious "alternative scenario", which, starting from the CO_2 target of the currently valid EU regulation of 95 CO_2/km in 2020, aims for an extrapolation to 70 CO_2/km in 2030 and to 45 CO_2/km in 2040. This scenario expects a massive change in environmental conditions, social and political upheavals of past trends, a noticeable change in consumer preferences, a rapid change in technological development, a significant shift in the fuel mix and a drastic reduction in the emissions target path. Achieving climates under this scenario comes with a higher cost to consumers.

In accordance to the DLR studies, Shell as well proposes a moderate "trend"-scenario with a CO_2 target of 70 gCO_2/km by 2040 and an ambitious "alternative"-scenario with

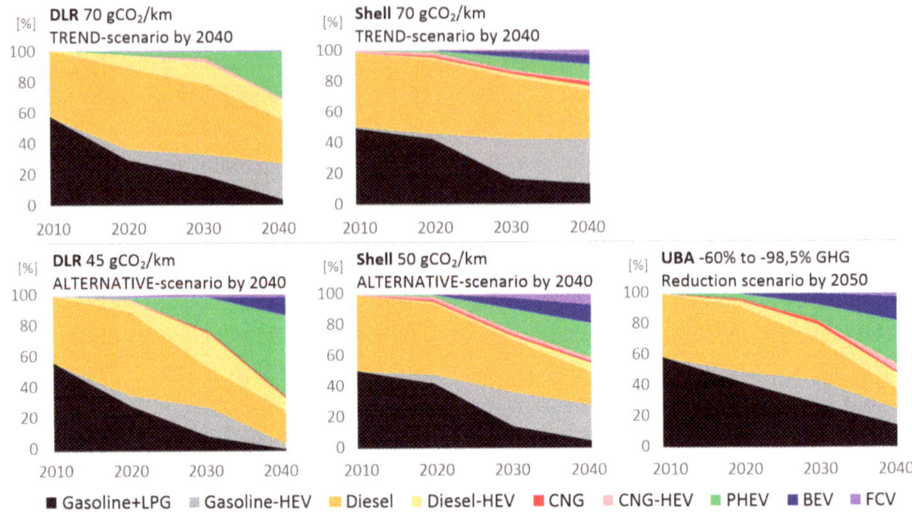

Fig. 4.2 Trend of new passenger car registrations along powertrain concepts in Germany until 2040 [2][26–28]

a target of 50 gCO_2/km in 2040. This makes them suitable for a one-to-one comparison (Fig. 4.2).

The "Climate Protection Contribution of Transport until 2050" by the Federal Environment Agency is the third study used here for comparison reasons. It argues over the commitment of the Federal Government to reduce greenhouse gases (GHG) across all sectors by 80%–95% by the year 2050. If this target is projected to the transport sector, it means a reduction in GHG in the range 60%–98% must be achieved. Transport in 2050 hence must be almost greenhouse gas neutral. Only higher cross-sectoral targets there will give more flexibility to the transportation sector [26–28].

In both alternative scenarios of DLR and Shell as well as the UBA scenarios, hybrid concepts (HEVs) bring about significant changes representing a combination of electric and conventional propulsion. Due to the relatively simple convertibility of conventional drives to 48V mild hybrids, these concepts will be a good option for an interim solution in the short term until 2025. The possibilities of $P_0 - P_4$ topologies allow a high degree of flexibility depending on the base engine concept [29].

In 2030, hybrid vehicles (mild hybrids and full hybrids) alone will no longer be sufficient to reduce the fleet's average CO_2 emissions sufficiently. This means that the proportion of the distance traveled with electricity will have to increase, which can be achieved by plug-in hybrids (PHEVs) and Range Extender Electric Vehicles (REEV).

[2] LPG=liquified petroleum gas, HEV=hybrid electric vehicle, CNG=compressed natural gas, PHEV=plug-in hybrid electric vehicle, BEV=battery electric vehicle, FCV=fuel cell vehicle, GHG=green house gas.

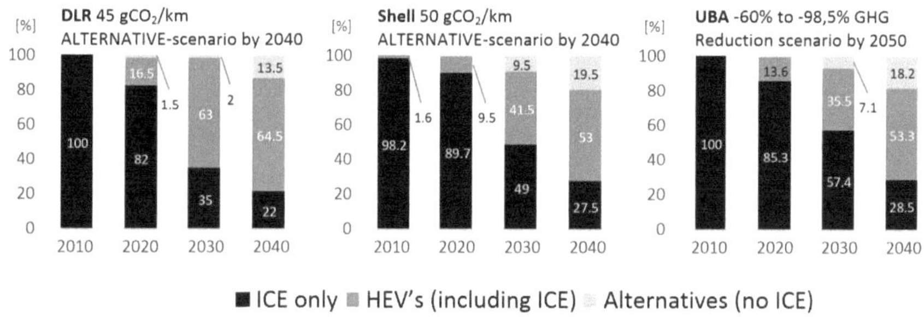

Fig. 4.3 Trend of the combustion engine as a proportion of all new car registrations [26–28]

The process of electrification will continue to be expanded until 2040. The PHEV share and the REEV share will clearly gain the upper hand compared to fully hybrid concepts. Pure battery-powered vehicles (BEVs) will continue to grow. At this point in time, the hydrogen fuel cell (FCV) continues to play a subordinate role in the market.

One question that still needs to be answered when it comes to the future of drive concepts is the role that the combustion engine will play. From the three ambitious scenarios of the presented new car fleet, a distinction is made between concepts relevant to the combustion engine and those not relevant to the engine; see Fig. 4.3. According to the DLR study, 86.5% of all powertrains will still have a combustion engine in 2040. The Shell study forecasts this proportion at 80.5% and the UBA study at 81.8%. What emerges from all three studies is that even with a drastic change in technology as a result of the climate targets, the combustion engine will keep significantly the upper hand even in 2040.

Vehicle stock

The new vehicles that are added in Germany every year make up about 7% of the total vehicle stock. According to the Federal Motor Vehicle and Transport Authority of Germany, the average age of all registered passenger cars in January 1st, 2019, was 9.5 years. The development of the vehicle stock is therefore lagging behind new registrations by this margin.

The car stock is derived from demographic and socio-economic developments and is determined by age- and gender-specific motorization concepts and age-specific car performance requirements. Socio-demographic factors and spatial settlement structures will also affect the demand for mobility as in motorization performance. Hence, these factors will have a great influence on fuel demand and the trend for the car stock. In addition, income and subsidizing effects of specific driving concepts and fuels can have a considerable impact on the automotive landscape [27].

According to the presented studies, the following graph shows the stock development until 2040. In the year 2040, according to the DLR study, 94.5% of combustion engines will still be operating on German roads. Shell and UBA both forecast this value at 89.5% (Fig. 4.4).

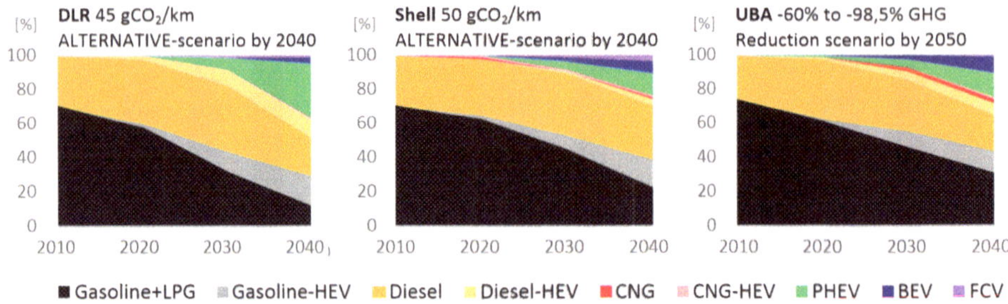

Fig. 4.4 Trend of passenger car stock along driving concepts in Germany until 2040 [26–28]

How do we prepare for these scenarios?

As a result of the targeted and versatile driving concepts, the development costs of automobile manufacturers will increase dramatically over the next years. Combustion engines have already reached a high potential regarding their efficiency and thus a minimum of CO_2 emissions. Further attempts at optimization are now only bringing marginal improvements at the same or even higher investment volume—but this does not mean that they are so easily replaceable. One thing, however, is certain—major investment trends for the development of internal combustion engines cannot continue as they have in the past decades. An investment basis resulting from more versatile driving concepts will only be profitable in the future if significantly leaner development processes and more efficient methods replace the current ones.

The right tools are available at the right time: artificial intelligence (AI). AI is the science of intelligent agents combined with the power of neural networks that primarily computer scientists have dealt with over the past decades. Their versatile fields of application have accordingly settled in the world of IT and have significantly advanced the digitization process since the1970 s. In the meantime, the concepts of AI have become versatile, reliable and highly efficient. How the topics of a digital IT sector can be transferred to industrial fields will be presented in the following chapters by means of work processes, concepts, methods and application examples.

4.2 The History of Simulation

Simulation as a development tool has significantly influenced the progress of technology since the1970 s. Irrespective of its field of application, it has been able to identify new optimization potentials, to compare and evaluate various system variants in a cost-effective and time-efficient way. It has also helped to define uniform development processes in a new way.

Only a few decades ago, powertrain development was test-based or prototype-based. This form of development was purely hardware-oriented and caused high costs over long periods. Each individual and new development stage that consists of compound systems (such as the powertrain, which is a compound of numerous subsystems) could only be evaluated in its entirety once a finished prototype was designed. On the basis of this working principle, further development steps or improving measures could only be achieved on the basis of intensive observations, combined with a high level of experience and sound knowledge of theoretical fundamentals.

Nowadays, the manufacturing of industrial products, especially such as powertrains, is almost entirely virtual and model-based. From the first concept paper to the completion of a prototype, powertrains can be developed entirely virtually, so that development costs today have shifted almost entirely from a hardware-based to a software-based process. Almost 50 years ago, it would have been impossible to foresee this progress and today's development structures. Likewise, it is not easy for us to estimate how things will work 50 years from now. However, one thing is beyond question: What helps to anticipate the future is to understand the past and the present. According to this principle, it becomes interesting to deal with the historical aspect of simulation, to what extent it has taken not only the automotive industry forward but also the world.

Figure 4.5 shows the result of a study in which all simulation software products (tools) developed on the market since the 1970 s are listed numerically. The focus of the study rests on the product levels that are equally relevant for the automotive area, here subdivided into 3D-CFD, 1D-CFD/0D-CFD, elastohydrodynamics (EHD), chemistry/kinetics, FEM, multi-body simulation (MBS), acoustics/NVH and kinematics/chassis. Only conventionally available products are included in this illustration. The market entry of a software product is rated +1 and a market exit is rated −1. The graphic shows how the trend of the respective product level has developed since the beginning of the conventionalization of simulation from the early 1970 s until today.

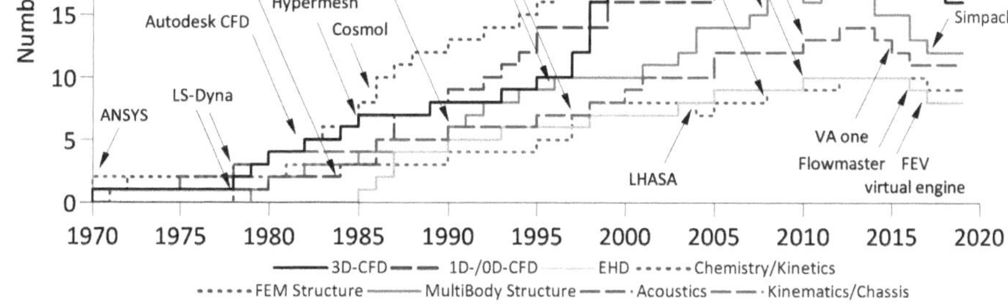

Fig. 4.5 History of conventional simulation tools for general applications

Until the end of the seventies, there are only a few products available. In the early eighties, software vendors had understood the potential of simulation-based development and correctly assessed a progressive increase in market demand, which led to a steep rise in supply. It is interesting to observe that despite individual delays, all product levels increased homogeneously, which indicates that a uniform implementation of all products into the complete value creation chain was recognized by customers and strived early.

In the 1990s and 2000s, the rapid increase took on. While software tools for chemistry/kinetics or EHD reached saturation early on, from the fact that a basic complexity could be achieved with very high predictive power, doors opened up for a deeper development potential for CFD (3D/1D/0D), FEM and MBS, so that product developers remained committed to satisfying the continuing demand for tools with higher precision and more complex requirements.

Around the year 2010, all product levels of the simulation were running into saturation and in the period between 2010 and 2015 a clear backward trend can be seen in all product levels. Fig. 4.6 allows a closer look at the simulation tools that are exclusively intended for the automotive market or at least target their core business there. The overall trend is almost identical. On a closer look, differences can be seen on the different levels between FEM and CFD (0D/1D/3D) as they grew closer together, which also underlines their capability for interactions.

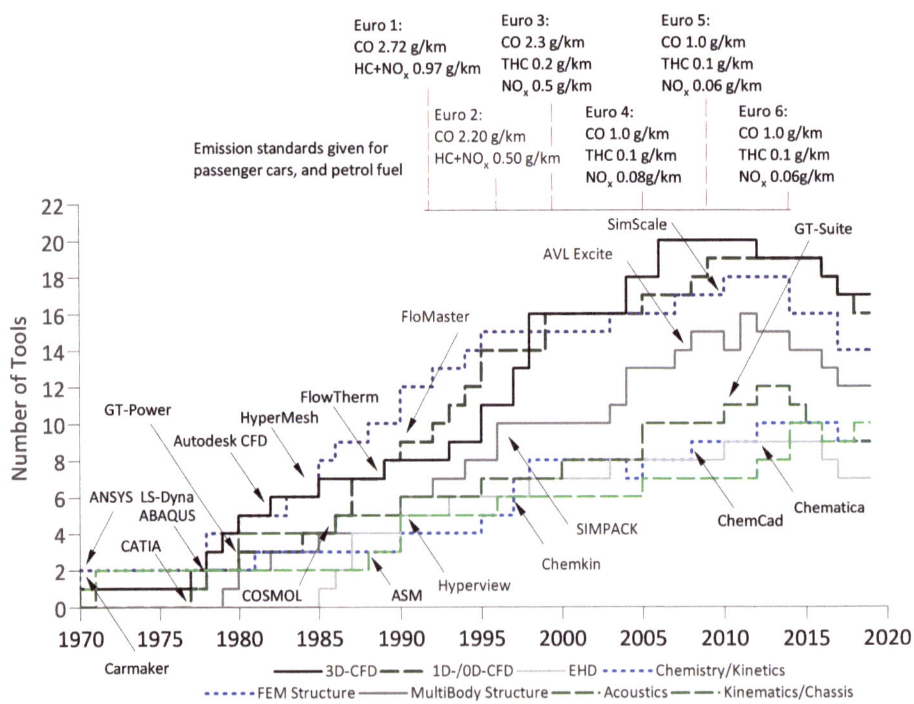

Fig. 4.6 History of conventional simulation tools for the automotive industry

Since 2013, the trend has been declining at all levels for which the following reasons are comprised.

1. **Centralization of development units and standardization of software tools**
Model-based development costs have increased significantly over the last decade. On the one hand, simulation software in general gains in complexity through continuous development, which leads to an increase in license costs. On the other hand, car manufacturers tend to use a high variety of software products, which is because usually different departments within a company often use different software products for historical reasons, which in principle cover the same product level. For this reason especially in recent years, it has been observed that companies have gradually centralized their pre-development units and standardized their software in order to save costs at this level. Nevertheless, the high performance of tools generates secondary costs, as they require centralized computing architectures and cluster systems. Last but not the least, there are high personnel costs for experts who have to bring or learn specific skills for the application of these software products.

2. **Change in the market**
Another reason why conventional simulation software is becoming less attractive for developers is that a declining sales market is expected in the future. The potential of the combustion engine with regard to improving efficiency is now largely saturated. The transition to electromobility and fuel cells does not demand the current variety of tools. As an example, structural calculation software (FEM/MBS) for a purely electric or fuel-cell-powered drive concept is losing relevance, as material stresses and loads play a much smaller role than before.

3. **A highly competitive situation**
The competition on the market for simulation software products has increased massively over the last three decades. Some products are well established within their level, others less. This has caused a strong economic gap, so that market-dominant suppliers repeatedly took over smaller and weaker products. Nowadays, it is almost impossible for small software suppliers to bring new products to the market, as they are under enormous competition and once established process chains in development units of car manufacturers can hardly be broken through advertising and marketing measures even at a great expense. These facts additionally contribute to the fact that the absolute number of products is decreasing.

Figure 4.7 summarizes the cumulated numbers of software products for general applications and specifically for the automotive sector. Looking at the emission limits according to the Euro legislation, from the introduction of the Euro 1 standard in 1992 until the

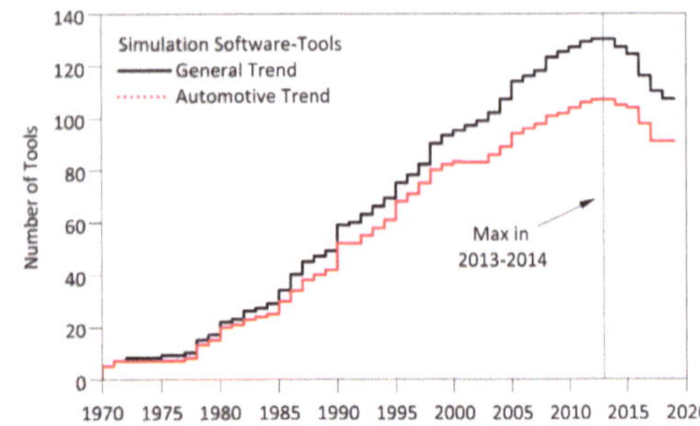

Fig. 4.7 Development trend of all simulation software products

introduction of the Euro 6 standard in 2014, the permitted CO content was reduced from 2.72 g/km to 1 g/km and the HC – NO_x limits from a total of 0.97 g/km to 0.1 g/km HC and 0.06 g/km NO_x. Including the Euro 5 standard, vehicle registrations were based on the NEDC driving cycle. Since then, the new WLTP[3] is being used as the basis for legislation; see Fig. 4.8 (left). China, which has the highest population density in the world, follows the EU's measures with its emission regulations. Eight years later in 2000, the first China 1 standard was announced. Since then, the country has caught up considerably with Europe and is now taking the lead with the early definition for the China 6b standard, which is comparable with the Euro 7 standard. Compared to the current China 5 standard, the CO and HC content of 1 g/km and 0.1 g/km, respectively, will be reduced by a further 50% and nitrogen oxides by a further 42%.

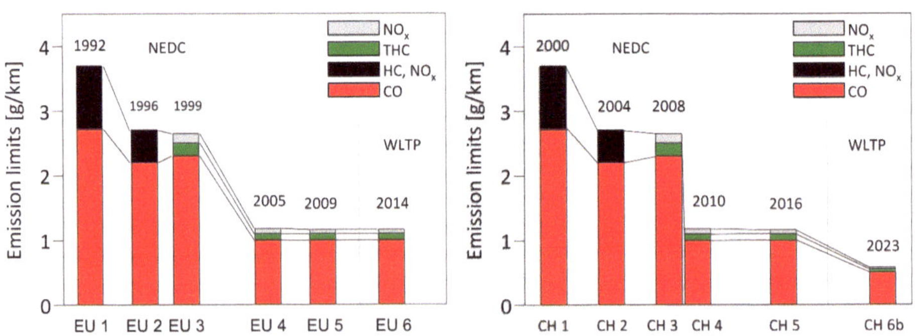

Fig. 4.8 Development trend of emission limits since the introduction of emission standards for the EU (left) and China (right) [30]

[3] Worldwide Harmonized Light Vehicles Test Procedure.

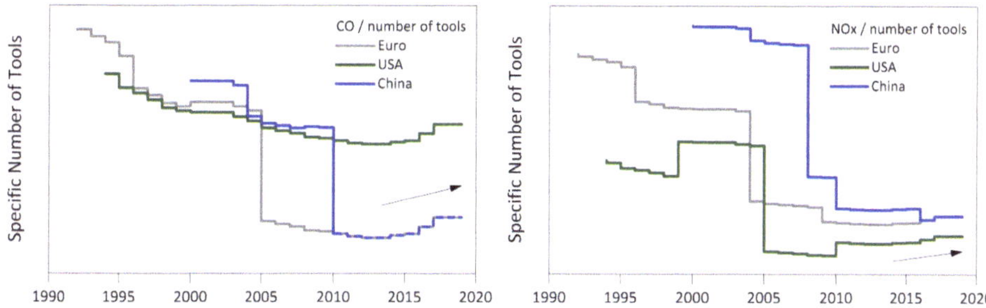

Fig. 4.9 Specific CO_2 (left) and NO_x (right) fleet development trend since the introduction of emission standards

When putting the market trend of the simulation tools in relation to the emission limits (here for CO and NO_x), one obtains a specific representation according to Fig. 4.9. The units [CO/number of tools] (left) and [NOx/number of tools] (right) give an indication of the extent to which simulation tools have helped us to reduce emissions using computer-aided and model-based methods in the context of powertrain development since the introduction of emission limits. Both graphs show how the specific value has been on a downward trend since the early 1990 s. For all three legislations (Euro, US Ulev[4] and China), an absolute minimum is reached around 2013. After that, the specific emission values have started rising again.

To show the same specific trend for the CO_2 development, the number of the simulation tools is put into relation with the CO_2 limits in Fig. 4.10. Also here, it is evident that the specific CO_2 value reaches its minimum between 2014 and 2015 and has been increasing again since then.

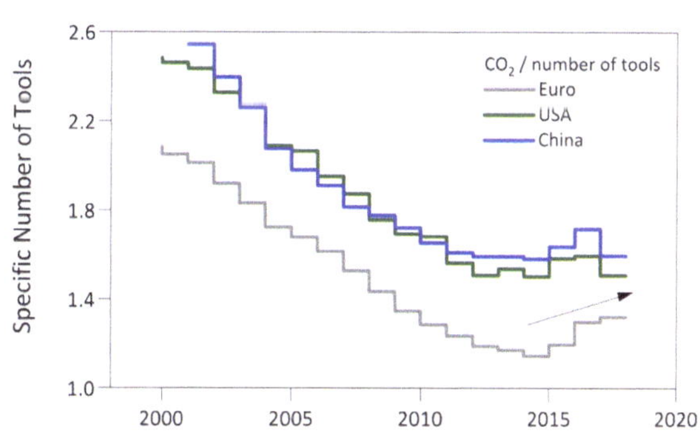

Fig. 4.10 Specific CO_2 fleet development trend since the introduction of CO_2 limits

[4] Ultra low emission vehicle.

4.3 Development Processes and Scenarios of Simulation with Big Data

As a result of the rapid development of data storages, starting with the simple floppy disk from the mid-1970s with a capacity of about 80 kilobytes and the market launch of recordable compact disks in the early 80s with several hundred megabytes suitable for series production, in the early phase of the digitization process it was well understood that the technological possibilities of storage and the availability of complex data volumes would strike new ways with inconceivable potentials. The term "Big Data" was first popularized in this context in 1990 by a US-American computer scientist named John Mashey.

Today, big data no longer stands for technological progress in handling large amounts of data solely, but for a breakthrough in new development methods for modern companies. It does not matter in which segment a company is positioned—the methods are universally valid and can be used in any field. However, the preparation (pre-processing) can vary greatly and play a decisive role in a successful implementation (Fig. 4.11).

From the possibility of generating large and indefinable amounts of data and making them storable, the real challenge arises, namely the systematic handling of these data. Big data is a far-reaching term and can be characterized by four properties for every conceivable amount of data:

- Data Volume describes the size of a data set.
- Velocity describes the speed at which data characteristics or the amount of data itself changes.
- Variety describes the expandability of a data set and the variability of a data source.
- Veracity describes a degree of average noise that naturally comes with real data.

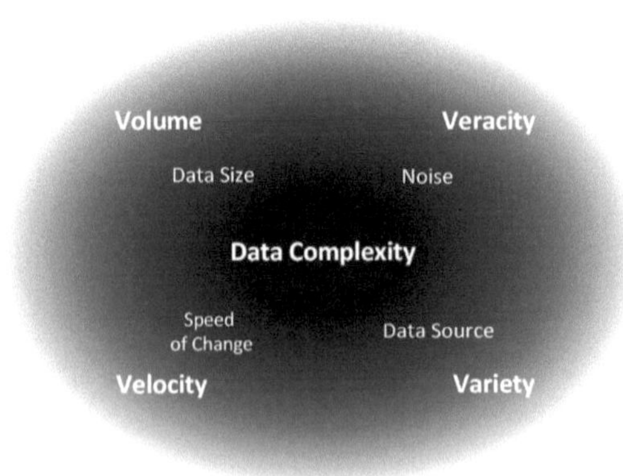

Fig. 4.11 Characteristics of data (Big Data) [31]

How can data be generated?

Until about the end of the1960 s, when simulation tools were not yet firmly anchored in the processes of powertrain development, decisions for any development were purely taken based on testing results. Every development step was under close observation and without modern tools, as they are used today, directions and progress could only be achieved through a solid experience and a constant elaboration of theoretical foundations. The design unit, production and test bench operation were in a much closer interaction than today. This was necessary since the purely test-based development of systems and subsystems were requiring the production of new prototypes for every single development step. Only a validation of a manufactured system could provide a statement as to whether and what qualitative gain this development step brought about.

Figure 4.12 shows a general process chain for the development of a powertrain in a simplified form, as it was done before the age of simulation. A reference motor with the prototype designation P_0 is at the beginning of a new development stage. On the basis of defined targets (specifications), subsystems (component level) are optimized (R), adapted in terms of design (D) and manufactured (M). This can relate to cylinder and piston geometries, manifold inlet and outlet geometries, materials for higher load capacity, better thermal management or for weight reduction of mechanical components, ignition devices, the crank mechanism, valve train or any other form of system. The next step is to produce a modified

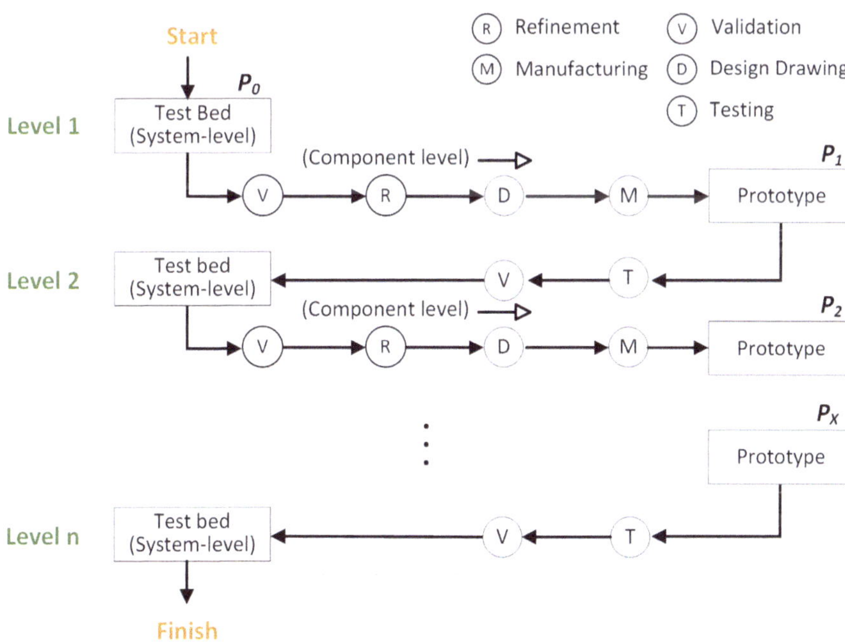

Fig. 4.12 Prototype-oriented/test-based development process of powertrains up to the1970 s

system with the prototype designation P_1. In a closing loop, the new prototype is now tested (T)—a qualitative evaluation can be made according to a new validation loop (V). This process loop is repeated many times until the target of the specification sheet is reached by the prototype variant P_X and all subsystems are merged into a complete system (system level).

Nowadays, development departments of all automotive manufacturers strongly integrate simulation into their development routines. The theoretical foundations of some of the topics were presented in Chapter 3.

Since the beginning of the commercialization of simulation tools and their use in corresponding development phases of powertrains, such as concept design, material selection, construction, pre-application and application (see Fig. 3.1), development processes have changed dramatically. First and foremost, simulation had produced serious potentials that could not have been localized and implemented at comparable speed on a purely test-based approach. In addition, the avoidance of permanent production of prototypes resulted in significant cost savings of hardware.

A new process chain allows the virtual development of complete systems (system level) along continuous development phases (level) in a purely model-based approach. On the component level, the development comprises the chain of virtual calibration (C), optimization (O) and computer-aided design (D) of model levels ($M_1, M_2, \ldots, M_n$) and the subsequent data generation (DG) for the calibration of electronic control units. Only after the entire process has been completed the first prototype P_1 is produced, which immediately fulfills a large part of all the required specifications. Manufacturing is thus shifted to the very last step of the process chain (see Fig. 4.13).

Simulation is stronger than ever

Over the past decades, university research institutes, research and development units of automobile manufacturers as well as suppliers and external service providers have been intensively engaged in the field of simulation worldwide. This has resulted in a large number of publications that have gradually continued the success of virtual development. What can be designed surprisingly well with simulation approaches are methods and methodologies for advanced processes. Methods are single solution concepts, methodologies are the composition and process sequence of many single methods in a closed development framework.

Methodologies have now become overarching sub-disciplines in the field of virtual driving concept development, holding structures and strategies inside development units together representing the uniqueness of corporate cultures or even microcultures on a department level. Since all automotive manufacturers use the same or similar simulation tools by today, it is even more important to distinguish oneself by individual methodologies and create a unique development DNA.

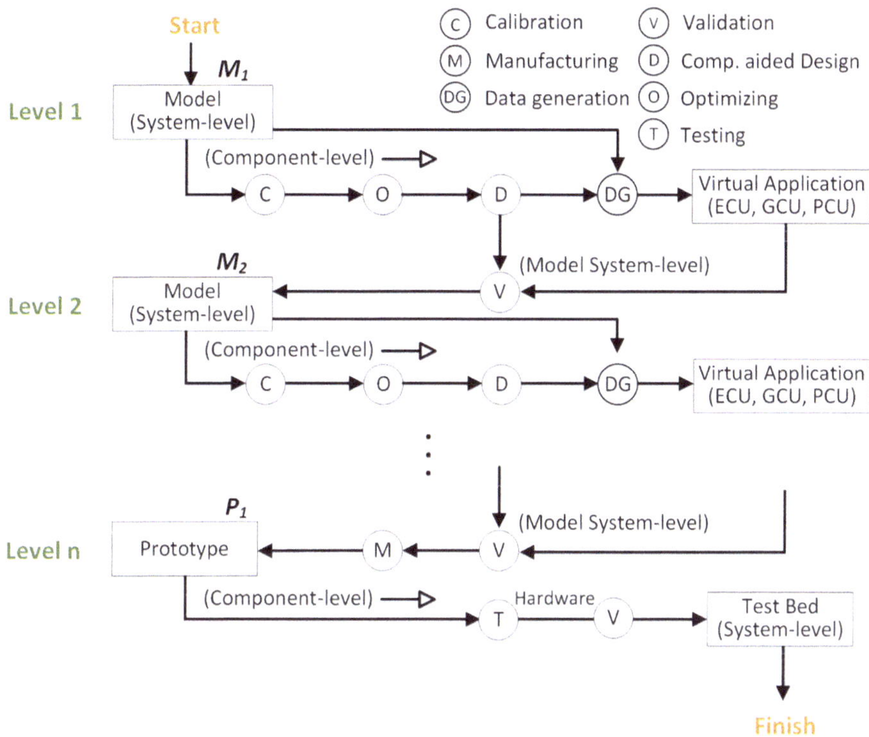

Fig. 4.13 Simulation-oriented development process of powertrains today[5]

Figure 4.14 (left) shows schematically how the quality and reliability of simulation tools have changed over time. As a result of the continuous development of model approaches, simulation tools are extremely reliable in terms of predictability and are close to reality, at least as far as qualitative trends are concerned.

In the past, simulation tools were used independent from each other on an individual demand. Therefore, it was essential to have realistic boundary conditions in order to make predictions of high quality. The graph on the right shows how simulation among powertrain development has evolved from an initially processless to a process-integrated tool since the 1970 s. Accordingly, the trend of test-based methodologies has declined. Today, test-based methodology plays a higher role at the end of the development process chain in the context of system validation.

As explained at the beginning of this chapter, big data provides the basis for the development of models that make use of concepts of artificial intelligence. Without the existence of large and, above all, diverse data sets, the highly effective algorithms of AI cannot be used. In order to make use of AI, the idea of generating data sets on real test benches would be an obvious one. However, this process would be extremely cost-intensive and not target-

[5] ECU=Electronic control unit, GCU=Gearbox control unit, PCU=Power electronic control unit.

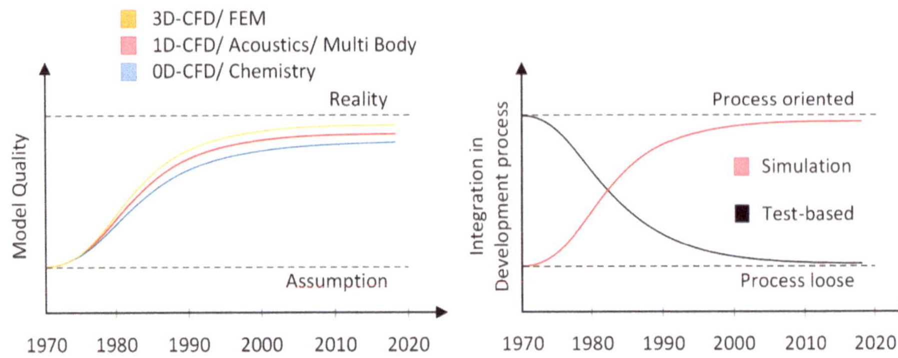

Fig. 4.14 Reliability of simulation (left), Historical change of development processes (right)

oriented, since real systems only generate data within the boundaries of systemic plausibility. An algorithm based on this data would be strongly restricted in its freedom and generate solutions which from the point of complexity cannot go beyond already known. What we expect from AI is not only new results, but above all new solution paths that leave the limits of systemic and human plausibility to create unexpected and unlimited solutions.

If we now try to implement a resource-oriented thought based on what is available to us, we would use simulation tools and models to generate massive amounts of data on a purely virtual basis in a short time. A possible basis for data generation is provided by the Design of Experiment (DoE). Fig. 4.15 compares both approaches, namely measurement-related data and simulation-related data generation. In principle, it is possible to use any simulation software as a data generator.

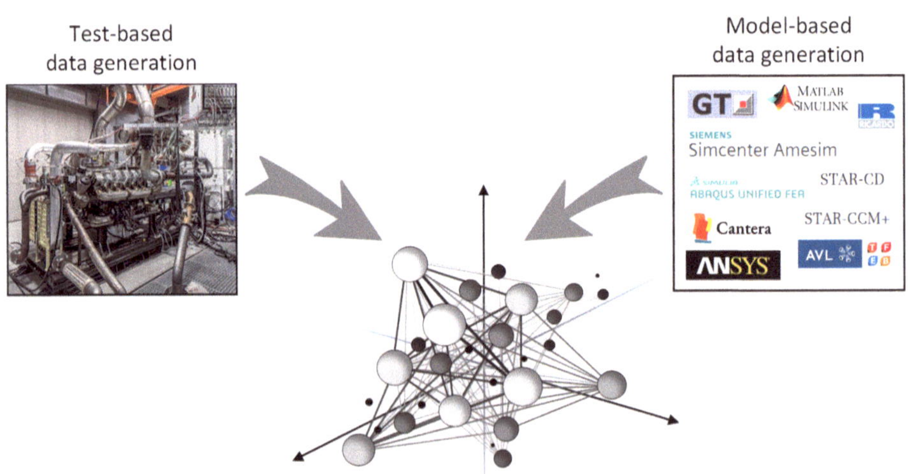

Fig. 4.15 Generating big data through real data (left) or through simulation software (right)

Advantages of AI-based powertrain development
The advantages that AI-based concepts can provide for the development of powertrains are obvious. While calculation in conventional computing systems are ordered serially, and lead to a lower data processing speed, an AI architecture processes in parallel and therefore gains an exponential speed advantage. Especially the associative operation allows finding creative and complex solutions to problems, which a normal computer code would not be able to solve due to its address-based principles. This results in many types of problems that could not, or only with great effort, be converted into an algorithmic form so that they could be solved by a computer. Also, unlike the processes of an AI system, conventional algorithms do not undergo a learning process and are not adaptive for recurring problems.

Figure 4.16 shows how AI calculation processes can be classified in terms of accuracy and computing time compared to classical methods of simulation. In general, the levels of simulation can be divided into differential equation methods and signal-based methods.

3D-CFD (see Sect. 3.1) presents the highest level of detail using the method of partial differential equations (PDE) for solving a problem. Solvers are computing both time discretely and space discretely for each individual node of a grid (mesh), and this is done several times per time step, depending on the order of the solver. 0D represents the lowest PDE-based level for which a spatial resolution is not executed (see Sect. 3.1.3).

The signal-based level is divided into white-box, grey-box and black-box models. White-box models are physical models that represent inputs and outputs based on theoretical (physical) relationships and are comprehensible for the user. Gray-box models, on the other hand, are hybrid models and known to combine a partial physical representation with additional

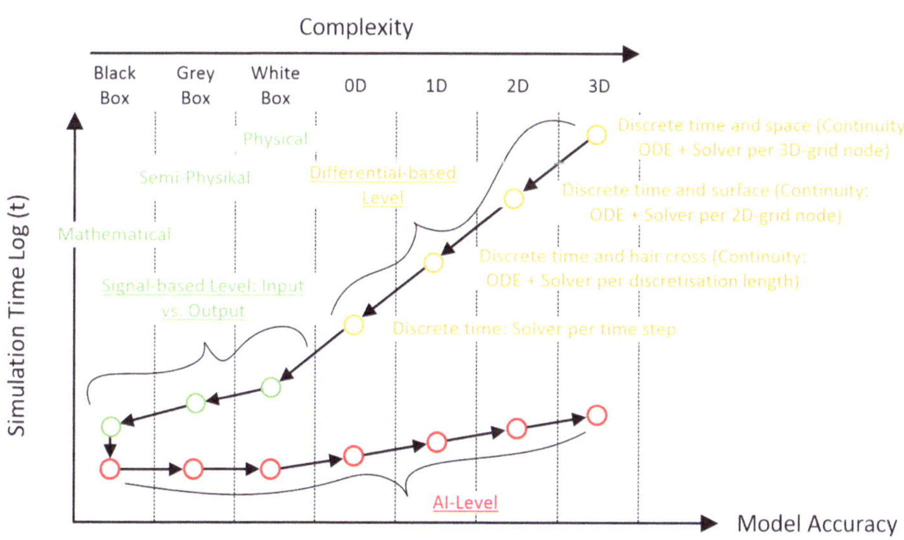

Fig. 4.16 Trade-off between level of detail and model accuracy: AI simulation compared to conventional simulation

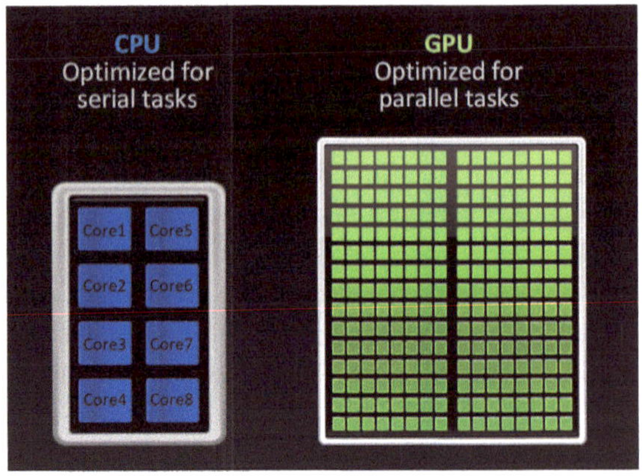

Fig. 4.17 Serial tasks versus parallel tasks

real data, which substitutes missing physical correlations through mathematical regressions. If no physical relations are known at all, so that correlations of inputs and outputs only have to be described by mathematical formulas, this can only be achieved by using large amounts of real data. The correlations remain intransparent for the user, which can be attributed to black-box models (Fig. 4.17).

While common central processing unit (CPU) chipsets in computers use multiple cores that focus on sequential processing of computational processes, a graphical processing unit (GPU) is suited for parallel processing. In contrast, it provides hundreds to thousands of smaller cores to process threads (or instructions) simultaneously. In combination with artificial intelligence and the parallel processing of neural networks, it becomes clear that GPUs are ideally suited for this purpose. The following graphic shows the result of a study in which the calculation speed of a CPU and a GPU are compared. For the benchmark, 4 neural networks were put to the test, with the increasing complexity of the architecture (Fig. 4.18). More details about network architectures are given in Sect. 5.5.4.

All PDE-based as well as signal-based layers can be simulated by AI. Hence, an AI level positions itself flexibly and, depending on the requirements, from simple to complex. When following the directions of the arrows, it becomes clear that even 3D problems at the highest detail level of simulation can be captured with AI, with a massive advantage in computing time.

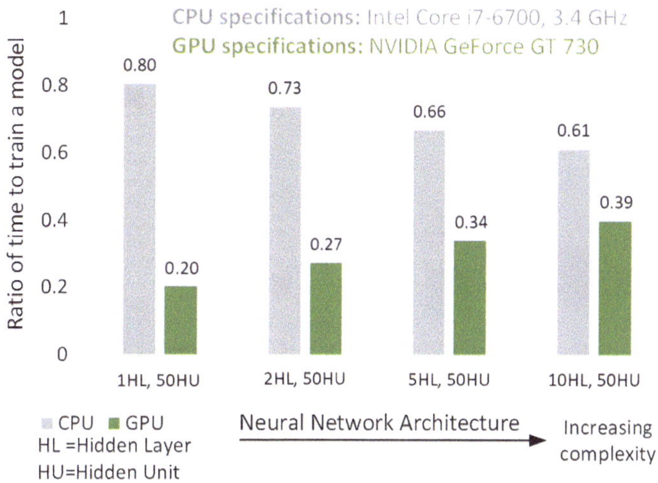

Fig. 4.18 Speed advantage GPU versus CPU

Exploiting the current trend toward new value creation processes
Operating speed is always directly related to costs. By applying AI model approaches, not only previous calculations but also methods and full methodologies can be integrated into existing development processes. This should enable not only a drastic increase in development speed but also a significantly leaner and more efficient process. In Sect. 4.2, the trend of the developed simulation tools until the1970 s was presented and discussed. The rapidly increasing demand for software until 2013/2014 (see Fig. 4.7), many of which are in competition with each other, has also confronted users with the difficult task of finding out which tools meet their requirements best and are integratable into their individual workflow. The consequences are clear. On the one hand, this leads to time-consuming and cost-intensive benchmarks. On the other hand, to high investment volumes such as for multiple licenses within identical application areas and high IT maintenance and network cost due to the diversity of softwares.

Following the idea, it becomes clear that the current trend together with the promising concepts of the AI interlocks ideally. A possible scenario that is presented here is that there will be a regression and reassortment of the software varieties. It is to be expected that software in the future will be limited to a core product that meets the requirements of the users in the broadest sense. In each discipline, whether 3D, 1D/0D, EHD, chemistry/kinetics, FEM, MBS, kinematics or acoustics, cost-saving measures can be taken by selecting the most efficient tool among a large variety for the respective needs and using it as a data generator from thereon. The development landscape will thus be composed of harmonious interaction of the strongest tool in its discipline (as a data generator), surrounded by many AI applications (AI gadgets) that use the input of virtual data. This rationalization process is referred to as gadgeting in the following (Fig. 4.19).

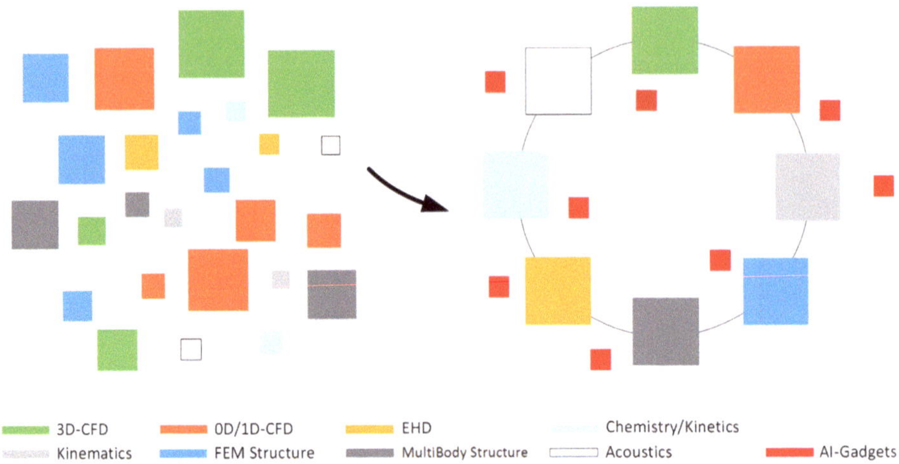

Fig. 4.19 Gadgeting process: Rationalization and reassortment of all simulation tools to the strongest tool in its respective discipline, extension of the development landscape with AI gadgets

In principle, this results in two interesting concept possibilities:

1. Partial concept: AI gadgets to support the calibration

A first concept possibility is to keep simulation tools as result- delivering systems. Additionally, they are used as data generators. The AI gadgets integrated into the structure have the function to set parameters of desired submodels in correlation with their initial results. In addition, the calibration effort (C) of models is automated so that a time-consuming para-

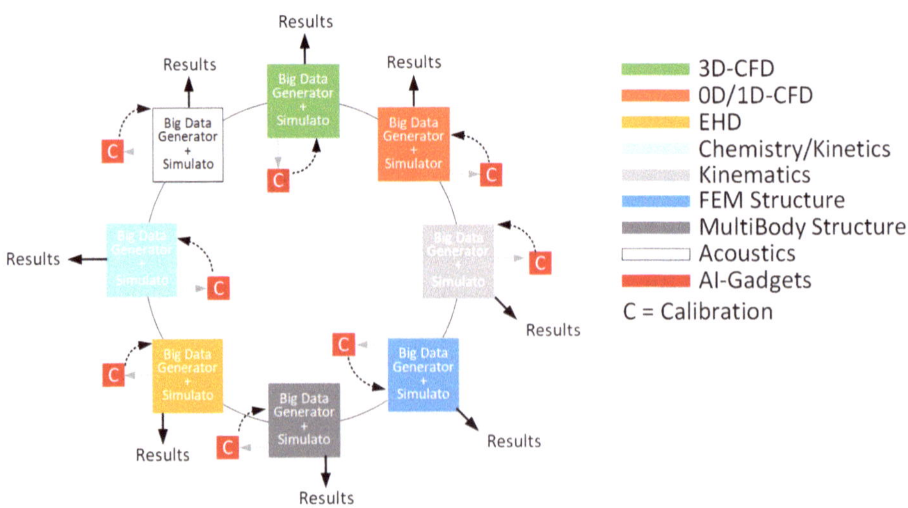

Fig. 4.20 Gadgeting sub-concept: AI gadgets to support the calibration

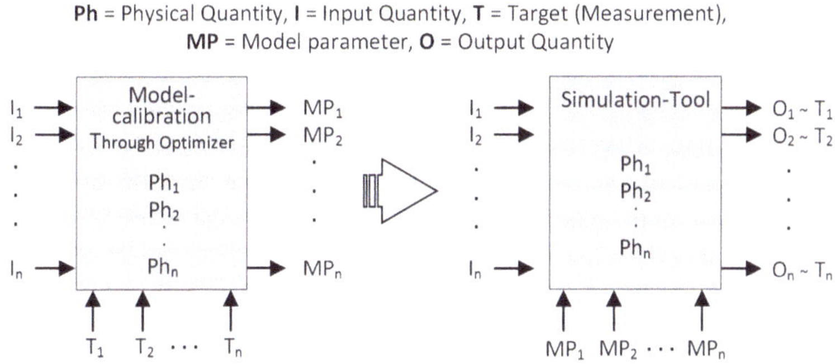

Fig. 4.21 Typical procedure of a model calibration (left) and applying a model as a data generator (right)

meter calibration to existing real data will not be necessary anymore. From there on, the correct parameterization is provided by AI gadgets. Fig. 4.20 shows how all systems interact in such a constellation.

To concretize the concept a bit more, it is shown in Fig. 4.21 how AI gadgets can be created to support model calibration. A typical two-step process chain is illustrated that is used to calibrate simulation models. In the first step (left), target values ($T_1, T_2, \ldots, T_n$) are specified, which usually originate from measured real data. By using optimizers, model parameters ($MP_1, MP_2, \ldots, MP_n$) are adjusted until the best combination is found which fulfills the target values at given input values ($I_1, I_2, \ldots, I_n$). In the second step, the results of the optimization (right graph) are used in the model, whereby it is now considered calibrated for the prediction of desired applications. After a tuning process, the resulting outputs should approximately match the target variables ($O_1 \sim T_1, O_2 \sim T_2, \ldots, O_n \sim T_n$).

The adaptation of a gadgeting process can be realized by a three-step procedure, (see Fig. 4.22). In the first step, parameters of the model ($MP_1, MP_2, \ldots, MP_n$) are varied cross-

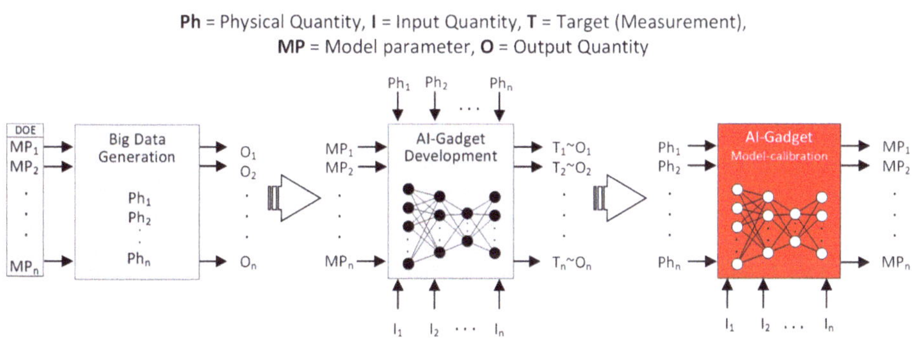

Fig. 4.22 Method for big data generation and creation of AI gadgets

wise over a DoE test space. The simulation tool acts as a data generator and generates result variables ($O_1, O_2, \ldots, O_n$) for each variation. In the second step, these variables are correlated with the physical input variables ($Ph_1, Ph_2, \ldots, Ph_n$), which are relevant for the considered model, and further input variables ($I_1, I_2, \ldots, I_n$), which are required by the model. By the approach of a suitable AI model, the aim is to generate the same target variables that corresponded to the model output in the first step. In the future, the AI model can be applied. Instead of performing a complete calibration again, the AI gadget is used, which initiates a reliable and fast prediction of the respective model parameters.

According to this idea, a modified process loop results. The delivery of results remains the responsibility of the simulation tools—the parameter calibration of all models, however, is done by the corresponding AI gadgets. If this procedure is integrated into the development, the following novel process loop results (Fig. 4.23).

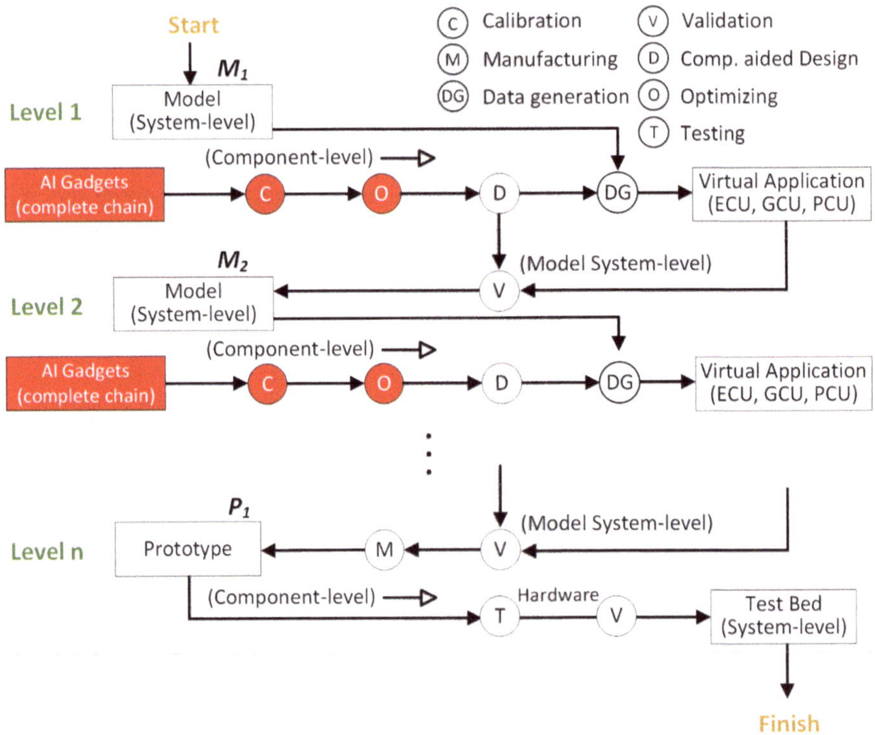

Fig. 4.23 Development of a prototype production: Gadgeting sub-concept

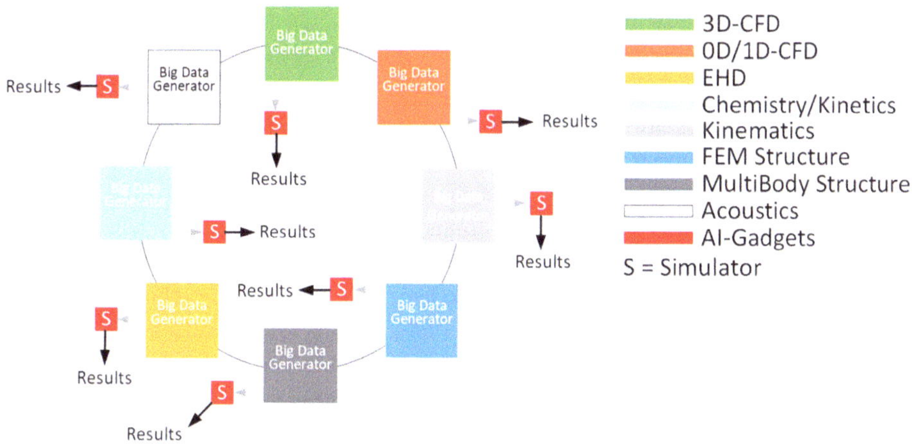

Fig. 4.24 Gadgeting full concept: AI gadgets as a result supplier

2. Full concept: AI gadgets as result supplier
The second and strongly extended process for using the lean structure of selected simulation tools is their pure use as data generators as shown in Fig. 4.24. For the generation of further results, the full concept is exclusively based on AI gadgets. Which powertrain concept is actually being developed plays a subordinate role.

A corresponding process chain within a development unit is illustrated in Fig. 4.25. In principle, it is conceivable that multiple AI gadgets can be interlocked and components can

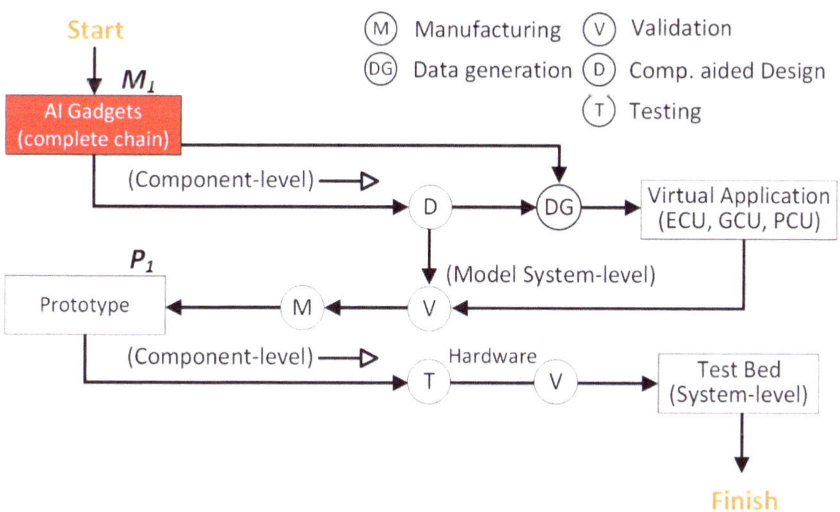

Fig. 4.25 Development process of a prototype production: Gadgeting full concept

be developed on a subsystem level in a single-step process. On the way to the first prototype (P_1), massive time saving is achieved. Finally, validation of the component level and the system level completes the development loop.

A number of topics are mentioned here, such as how gadgeting can be used in the development of combustion engines and so to concretize the novel process loop. Fig. 4.26 shows the most important submodels that are very typically performed using the means of simulation. Here, a boundary is drawn around the combustion chamber, which contains relevant submodels for the development of combustion processes. A typical loop within here encloses the topics (turbulence, charge exchange, ignition delay, combustion, wall heat losses, cyclic variations, knocking, raw emissions, exhaust aftertreatment and acoustics), here described as extensive effects. To be able to execute these models, boundary conditions are needed, which are represented by further submodels. These are characterized by intensive parameters.

In principle, it is possible to replace any individual submodel or even a group of submodels, which are usually determined physically, by the concepts of AI. Intensive parameters stand for quantities that have a direct and significant influence on the engine behavior — extensive effects, on the other hand, express the corresponding influence on a physical level.

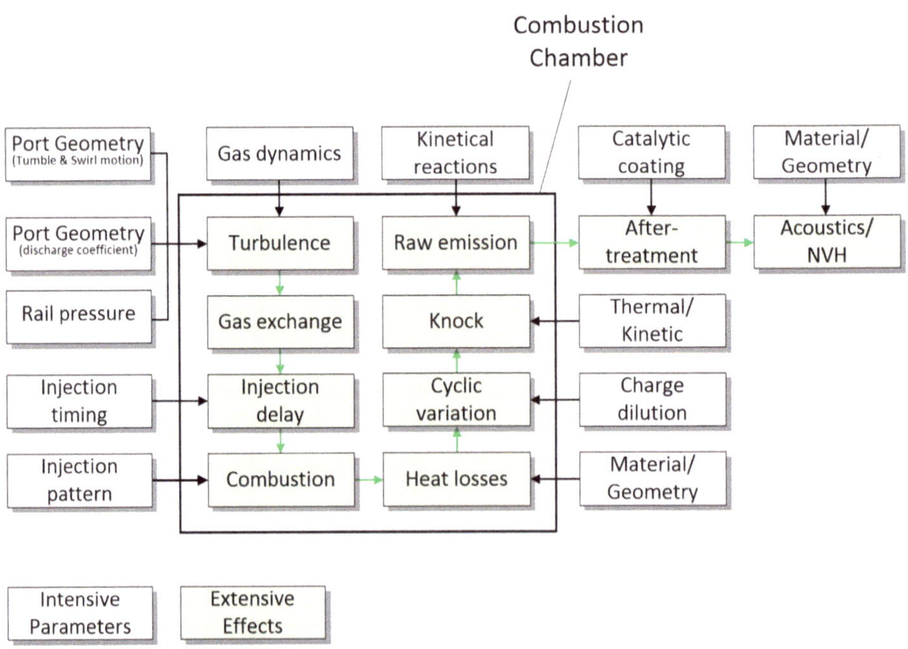

Fig. 4.26 Submodels in the development/simulation of combustion engines

During data generation (big data process), it is crucial which design of experiment (DoE) is selected to fill the design space. Additionally, each level of AI requires a different DoE method. There are basically four categories to be differentiated:

1. Space Filling Design
This experimental room design is suitable for both continuous (variable) and discrete (fixed) input parameters. The space filling design aims at filling the experimental space in such a way that all spaces are evenly covered, i.e. the distance between any two adjacent experimental points is maximized. In case of measurement outliers or areas of high sensitivity, this method is suitable to cover the experimental space evenly in the best possible way [32].

2. Robustness Design
This design variant is suitable for test spaces that require high robustness due to the interaction between controllable and uncontrollable noise. Control factors are used here, whose settings are made by the user or automatically and ensure that the noise of the test output is minimized.

3. Statistical Design
Statistical design of experiments provides an organized approach to generate data and optimize any process with multiple parameters. In a DoE approach, experiments can be run in random order while changing several variables simultaneously, rather than varying one parameter at a time while keeping all other parameters constant. The advantage of randomly selecting experiments is that each of them, as part of an overall population, has the same chance to contribute to the training of a model.

4. Optimal Design
Statistical designs such as the Full Factorial require an ideal and simple experimental setup, which is unsuitable for fulfilling multiple experimental objectives. The Optimal Design, on the other hand, is an approach that tries to take into account the entire observation space and to fully exploit the accuracy. It focuses on the minimization of a criterion related to either variance or other statistically relevant quantities. Two of the most common criteria are the D-criterion and the I-criterion. The D-criterion refers to the variance of factorial influences and the I-criterion to the accuracy of the prediction (Fig. 4.27).

Integration of AI-based development into lifecycle models
Lifecycle models set important quality and safety standards in the implementation of products within development processes. They serve as guidelines that support project monitoring and keep necessary development steps together so that they are not overlooked. Due to their phase-oriented basic framework, they provide decisive advantages in their application for the project management of product developments, not only for the project manager but also for the entire development team.

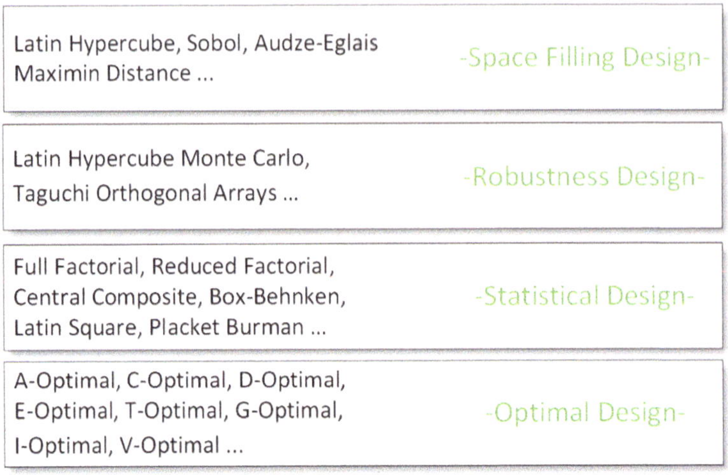

Fig. 4.27 DoE approaches for the creation of an experimental space

With the introduction of controlled development lifecycles, it could be proven that qualitative standards of product results could be raised in the medium and long terms. They demand a draft of system and software architectures in an upstream project phase and emphasize that all requirements for a system must be clearly defined before the start of an implementation. After the draft phase throughout the operational application, they serve as an orientation and contribute to realistic time and cost planning.

Especially in the automotive sector, the so-called V-lifecycle, which can be assigned to the classic lifecycle models, has become generally accepted over the last decades. Primarily in this sector, it is important to consider different components of products as well as software, mechanics, electronics or mechatronics as an integrated unit and not separately. The strict orientation of V-lifecycle models supports mutual links between different system levels (such as hardware and software) and emphasizes test, verification and validation phases to ensure the integrability to each new development stage.

The following diagram shows that the integration of AI can be very sensibly designed in the entire left half of the V. This includes the feasibility phase, the concept phase, design and implementation, starting at a system level all the way down to a component level (Fig. 4.28).

Every company has its own historically grown development DNA and individual product creation processes. In this respect, lifecycle models must be tailored to meet the internal, project-specific requirements. From a historical perspective, it can generally be seen that development NDAs between companies were much more different just a few decades ago than they are today. Due to international and open communication platforms as well as conferences, where internal development methods are disclosed, lifecycle models that hold together and standardize the core of complete development processes and, above all, a globalized employment culture where employees transfer professional experience from different

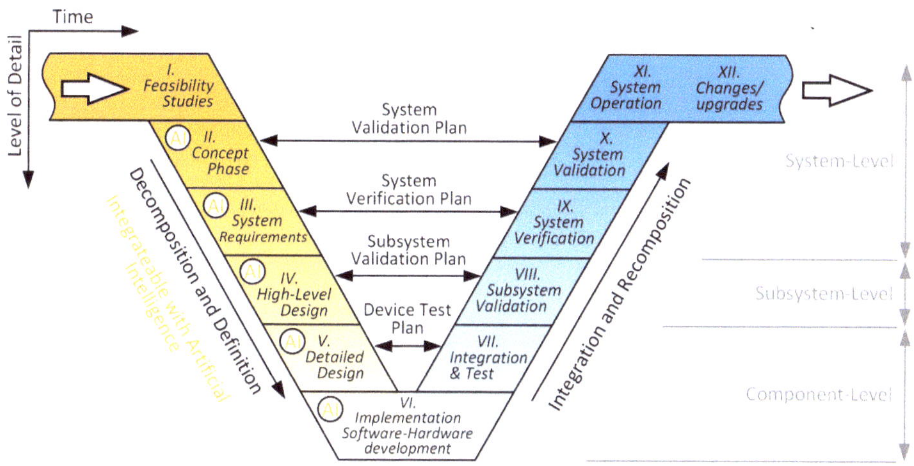

Fig. 4.28 Integration of AI into the V-development lifecycle for automotive applications [33]

companies to a new company, development cultures in the automotive industry are becoming more and more similar worldwide. In this respect, the transferability of lifecycle models has become easier and more flexible than ever.

Powertrain Development with Artificial Intelligence

5

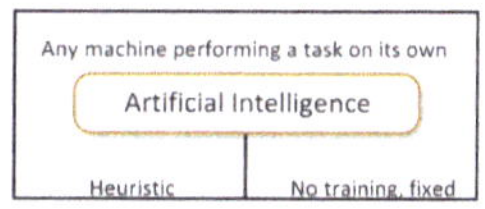

Artificial Intelligence (AI) is the science of transforming human learning, thinking and decision-making structures into mathematical models. Its goal is to enable machines to perform tasks intelligently and to detach them from an explicit path of programming a solution. The word AI was first coined by John McCarthy in 1956, who had the vision of giving computer systems a kind of decision-making power so that they work according to the model of human rationality. An intelligent system is defined as one that is able to perceive its environment and take measures to optimize any given situation on the basis of cognitive decisions. The variety of applications and imagination in the use of intelligent systems is great, and initially they did not primarily lie within the framework of technological developments but mainly in the medical field, such as neuroscience and psychology, as well as in disciplines of economics, linguistics and philosophy.

In the last few years, artificial intelligence has gained enormous importance. The availability of large amounts of data is the key basis for the application of AI. Since companies have been collecting large amounts of data of their business characteristics since the beginning of digitization, the majority of them are increasingly interested in using AI as a logical consequence. AI can help to decipher and interpret patterns within data and thus to extend development processes in an advanced way.

The original version of the chapter has been revised. A correction to this chapter can be found at https://doi.org/10.1007/978-3-662-63863-7_6

A. Mirfendreski, *Powertrain Development with Artificial Intelligence*, https://doi.org/10.1007/978-3-662-63863-7_5

5.1 Models of Artificial Intelligence

Artificial intelligence is a broad term and offers a wide range of possible applications. AI is often associated with the idea that machines or algorithms convey human thought patterns. It is important to realize that human thinking and its correctness can only be measured subjectively and does not necessarily have to correspond to rational thinking. An additional difficulty is that thought patterns or the solution of a thought process can trigger a behavior, whose correctness can only be judged subjectively. In this respect, a behavior as a result of a thought process can also be differentiated into a human and rational one.

Human thinking

The term thinking refers to all mental processes that combine imagination, memory, experience and a resulting process of cognition. Only the end product of thinking is consciously perceived as a result, rarely are the thought processes themselves accessible to one.

Thinking is a process that must generally be strictly separated from perception and intuition. Perception and intuition are not conceptual; in contrast, thoughts are propositional and tangible. If one examines the process of thinking, it can either be developed constructively or be triggered by an idea, a spontaneous feeling, a situation or by sensory impressions [34].

Thinking in detail is a current research topic in various disciplines. The complexity behind this lies in the construction of a thought structure, which can look very different depending on the discipline. In order to better understand the psychological, neural and biochemical mechanisms underlying a thought, brain researchers are studying the cellular level of thinking.

If a machine is attributed the ability to think humanly, then this refers in particular to the calculation process, which is comparable to the thinking process of a human. Thinking processes can be investigated and understood if they are related to the solution of a certain problem. In turn, solution paths can be manifold, which means that they have to be studied individually depending on the field of expertise and application in order to provide a tailor-made training for a machine. If a certain input-output pattern corresponds to that of a human, this is only the first proof that the thinking process could also correspond to that of a human.

A. Newell, J.C. Shaw and H. Simon developed the computer program "General Problem Solver" (G.P.S.) in 1959, which should be able to represent any form of problem definition by mathematical formulations. The focus for the developers was not to predict a correct solution for any given problem, but rather to design the solution path according to human logic in order to get to the same result. The interdisciplinary field of so-called cognitive science concentrates on bringing together computer models from AI and experimentally acquired techniques of psychology in order to determine precise test-oriented theories of the human mind [35].

Rational thinking

Making conscious decisions is an ability given to humans, which is only possible with the presence of imagination. An imagination of the effects of different or successive actions enables us to choose a preferred action from various options. In order to make such a decision, 1. a complex imagination in multidimensional levels and 2. target states are required, which make the effects of an action assessable. If an imagination corresponds to the actual effects and if, in addition, an effect is correctly measurable, it can be assumed that the original action is directly and functionally related to the solution. Both aspects are called epistemological and instrumental rationality.

Rational thinking is a skill that is not innate to humans, but must be learned and trained. Until then, our imagination seems to be afflicted with many shortcomings. It stands in contrast to irrational thinking. The ability to separate both thought patterns from each other is subject to a development process in our brain as it forms structures. The art of distinguishing conclusions that are correct in the majority of cases and therefore provide a good first estimation from cases that provide false results is called "heuristics". Since our brain learns through repetition, correct conclusions remain in our memory and overlay the cases in which we were wrong. On the other hand, if we reflect superficially, it is easy to overlook irrational thinking. This situation can lead to a situation in which one sees oneself in the right even if this is not the case. Human decision criteria are often based on experiences of many individual decisions from the past. Only the observation of past consequences, which have arisen from previous decisions, forms a sum of experiences, on the basis of which the following decision can be made. Thus, it can be said that human decisions are of statistical nature [36].

To know something means that it has been learned, and this in turn means that neuronal structures have been formed for it in the knowledge formation process. Knowledge can thus be directly related to complex functions in the brain. Unlike in an AI system, however, these functions are not activated by operators, but by biochemical processes and expressed through thinking. The reliability of biochemical processes can be equated to a probability thesis. According to this, probability is a property of the brain and a measure of the extent to which biomechanical processes can represent reality with the ability to abstract. Rational thinking is therefore also called thinking that is based on observable reality.

If one moves to the system level, systems are interpreted as rational if they deliver results "expected" by humans. Since the early sixties, programs have been developed in research, which are able to solve problems, which can be represented by logical notations.

Human and rational behavior

The British mathematician and logician Alan Turing is still considered one of the most influential theorists in computer science. In conjunction with his Turing machine, which was introduced in the mid-thirties, he succeeded in proving that machines can in principle succeed in reproducing complex relationships of reality on a mathematical basis. He postulated that cognitive processes can be reproduced as long as they can be broken down into algorithmic relationships. As a result of his considerations, in 1950 he presented the Turing test, which is still known today and is considered a recognized measuring method for proving artificial intelligence. The test model was defined in the written communication between a computer and a test person. The computer tries to confront the tester as a thinking individual. Only once if a communication is completed and the respondent was unable to determine whether his communication partner was a human or a machine was the test considered successful. To achieve this, the machine must be able to fulfill the following levels of abstraction [37]:

1. The processing of a natural language to understand the information it contains;
2. A knowledge compilation to store the exchanged information;
3. The automated inference to formulate appropriate answers from the stored information;
4. Machine learning to determine customized answers from the information.

Developers have succeeded in imitating human behavior over short periods of time, both in written language via so-called chatbots and speech generators. If the test person is informed, however, that the communication partner is a computer, they can easily expose it by asking specific questions. In order to model rational behavior algorithmically, so-called "agents" are used, which take on a monitoring function in a learning and training phase. The agent is assigned the task of grasping its environment, understanding input and disturbance variables in relation to output variables and autonomously adapting changes in the system to achieve the desired result. Furthermore, it has the task to imitate rationality by means of its control criteria. But there are situations, where no logical solutions to a problem exist. Especially in these cases, the agent is required to design reflexive solutions and extend its "rationality" to arrive at a solution that is not explainable via logic. Thus, the agent acts beyond the laws of logical thinking, because unlike humans it does not need to prove logical conclusions to its decisions. The agent is able to present results, which our mind could not achieve due to its limitation of the human thought structure.

Singularity

In connection with machines that have a certain degree of maturity to imitate human thinking and rational behavior, the term singularity is often used. Technological singularity describes theories that provide an outlook into the future. It is mainly understood to be a specific point in time that machines reach, from which they are able to develop autonomously and rapidly by means of artificial intelligence. This point in time is predicted as a possible, historical revolutionary event, behind which the future of mankind is no longer foreseeable. For researchers, the estimation of an exact point in time is currently difficult, but it is possible that it will come about unexpectedly. This expectation is based on the observation that technology and science have developed ever more rapidly since the beginning of mankind and are subject to an exponential speed of development, especially in the last century [38].

Singularity is often associated with the idea that human-like robots, also known as trans-humanoids, live next door to us humans and are far superior to us in all our characteristics, speed, strength, intelligence, availability and efficiency. However, according to the current state of knowledge, this idea is based more on science fiction than a realistic prognosis. The initial singularity will rather take place on an algorithmic and a process-optimizing level than on a robotics level. Already today, state-of-the-art and intelligent algorithms are used in almost all technological development levels to make a process more efficient. If these processes can be made even more reliable and, above all, more autonomous in the future, this will primarily mean that the working world of technology, economics, finance and banking and other sectors will continue to be subject to topological change. Accordingly, AI will creepingly rearrange the interplay between algorithmics and the labor market. The question of the extent to which jobs are endangered by the takeover of computers has long been under discussion. While this has not yet been proven to be a case, a reorganization of job topologies and a change in operative work are already underway. It should be possible to extrapolate this trend further for the coming decades. The following figure schematically depicts the point in time of the singularity on the timeline of human intelligence development (Fig. 5.1).

The beginnings of machine learning and the path to artificial intelligence go back to the 1950s. A multitude of individual achievements has paved the way for AI to become a powerful tool. The special feature of AI is now that it is available to all users through software. Already today, we are experiencing a remarkable increase in the number of new companies that use AI technologies. This leads us to expect that in the coming decade, economic and social digitalization processes will face a significant next wave. The timeline Table 5.1 provides an overview of the most important milestones of AI from its initial founding to the present.

AI is identified by economic studies as the key technology of the future. Many of these studies have already dealt with market potentials and their effects on the most diverse economic fields. A study commissioned by the German Federal Ministry of Economics and

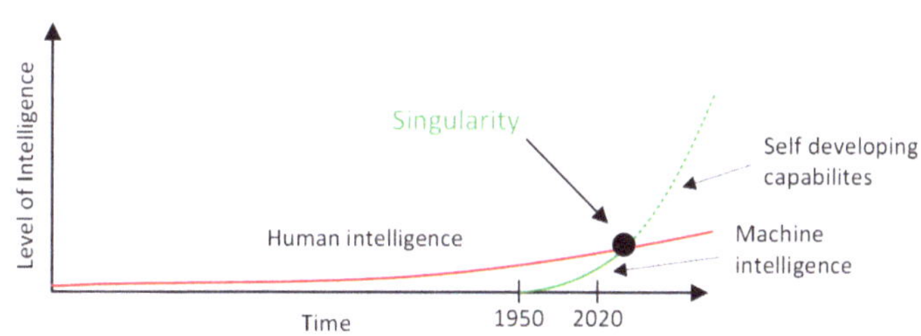

Fig. 5.1 Singularity [39]

Tab. 5.1 Milestones in artificial intelligence

1950	Alan Turing presents the Turing Test	1982	Dragon System introduces NUANCE, the development of the first commercial speech recognition system
1950	I. Asimov presents 3 laws of robotik	1986	Renaissance through neural networks by D. Rumelhart, G.E. Hinton and T.J. Sejnowski
1951	M. Minsky builds the first nuero computer called SNARC (Stoichastic Neural Analog Reinforcement Computer) with 40 synapses	1989	Carnegie Mellon University, Pittsburgh presents first autonomous vehicle based on neural networks
1955	First self-learning game „Checkers" is programmed by A.L. Samuel	1993	Computer POLLY gives tours on the seventh floor at MIT and interacts with visitors
1956	J. M. Carthy founds the word „Artificial Intelligence" as a research discipline on the scientific conference Dartmouth college, New Hampshire	1997	IBM DEEP BLUE beats the world chess champion G. Kasparov in chess
1956	A computer program named LOGIC THEORIST written by A. Newell, H.A. Simon and C. Shaw. It was the first program mimicing problem solving skills of a human being	1997	In Nagoya (Japan) the first robot world championship RoboCup is held with 37 participants
1959	MIT establishes an institute for AI	1999	Sony presents the first AI-based robot AIBO
1959	A. Newell, H.A. Simon and C. Shaw develop the computer program G.P.S (General Problem Solver) at the Carnegie Institute of Technology, Pennsylvania	2009	Stephen Wolframs develops the first semantic search engine called WolframAlpha
1961	The psychologist F. Rosenblatt implements perzeptron concept in the computer Mark I. Machine becomes adaptive through the trial error method	2009	Google builds the self-driving vehicle WAYMO
1961	First industrial robot „Ultimate" used in General Motors production line in Ewing Township, New Jersey	2010	Natural Language Generation (NLG) such as SARA are narrative AI tools presenting the ability to write documents
1966	The psychologist J. Weizenbaum develops the chatbot ELIZA, pretending to be a psychotherapist.	2010	Robust training of neural networs with multiple hidden levels becomes possible by G. Xavier and K. He
1968	First AI program for natural speech recognition named SHRDLU is presented by T. Winograd at MIT	2011	IBM supercomputer WATSON beats former Jeopardy champions
1969	Stanford Research Institute (SRI) introduces SHAKEY, the first intelligent locomotive	2015	Daimler presents the first autonomous truck on the A8 motorway
1970	Stanford University introduces MYCIN in LISP program, which analyzes blood infectious disease and suggests therapies	2015	Personal assistance systems such as Siri, Google Now, Alexa and Cortana are being commercialized
1971	The STANFORD-CART is presented as the first autonomous vehicle	2015	Open Source AI tools like TensorFlow, Azure, CloudML, Amazon AI, etc. are offered to users worldwide
1978	Facial action coding system (FACS) characterizes facial actions to express individual human emotions defined by P. Ekman and W.C. Freisen	2016	Google's Deepmind AlphaGo beats the reigning Go Champion Lee Sedol
1979	First AI expert system CADUCEUS for medical diagnosis specialized on internal medicine is presented	2016	NVIDIA introduces the Supercomputer for Deep Learning and AI
1980	LISP based computers are available on the market	2045	Singularity of robotics predicted. The AI's intelligence is superior to that of humanity and can evolve independently

Energy uses intensive surveys of companies in Germany to determine the extent to which AI technologies are already being used in the manufacturing sector by small and medium-sized enterprises (SMEs) and large enterprises (LEs) in their respective value chains. For a total of nine value-added chains, an average of 15% of SME and 25% of LE stated that they already use AI technologies at least to a small extent [40] (Fig. 5.2).

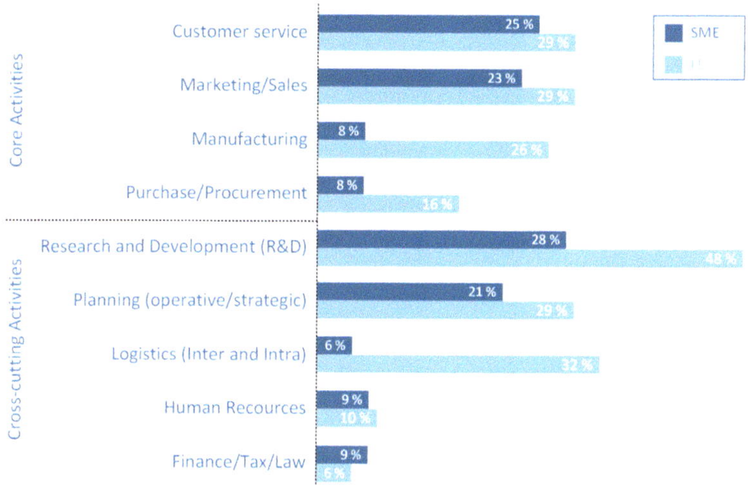

Fig. 5.2 Proportion of SME and LE that already use AI technologies, 2018 [40]

For a more comprehensive picture, the following diagram additionally considers the proportion of SME and LE cooperating with external service providers (Fig. 5.3).

A further representation permits the insight into the expectations of the SME and LE as a perspective outlook, specifically, how strongly it can be assumed that AI technologies will be used in individual processes of their value creation chains in the near future. The result clearly shows that all companies in the manufacturing or production sector foresee an increasing use of AI technologies in their processes (Fig. 5.4).

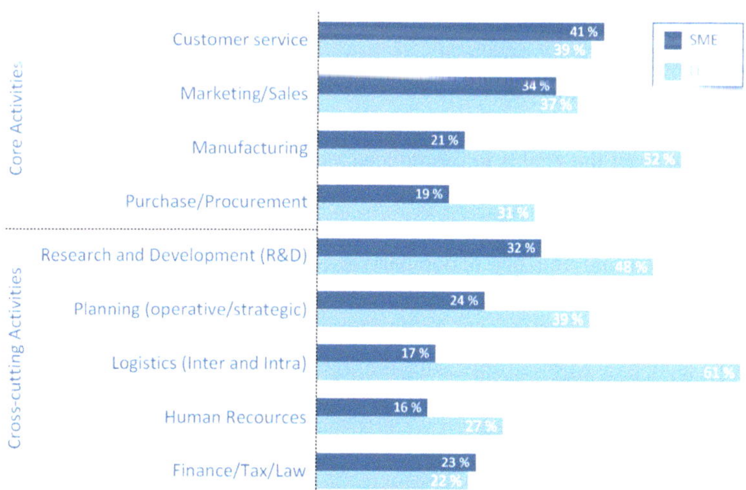

Fig. 5.3 Proportion of SME and LE which work with external AI providers, 2018 [40]

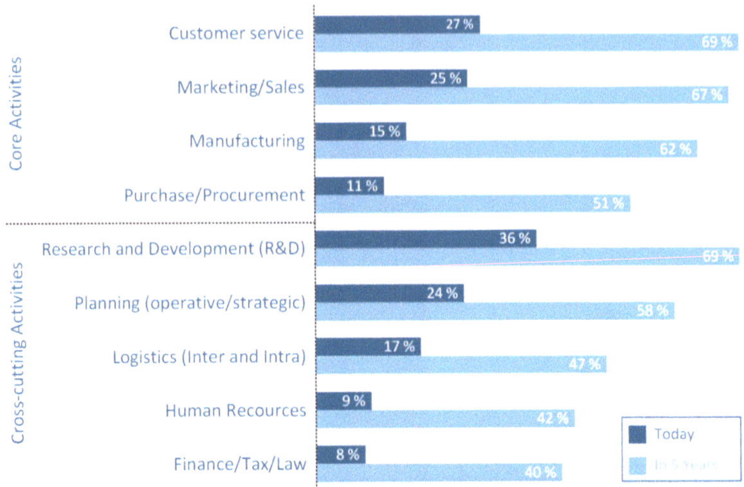

Fig. 5.4 Proportion of SME and LE that use AI technologies at least to a small extent today and probably in 5 years, 2018 [40]

5.2 Overview: Levels of AI

If models are discussed in general, a distinction is made between **deterministic** and **stochastic** models. If a recurring, identical input of parameters is followed exclusively by the same output, the model is defined and reproducible. This is a characteristic that applies to deterministic models. Here, the algorithm is clearly defined, i.e. all intermediate results that are generated are also identical. This is in contrast to non-deterministic models. As a rule, all realistic processes are not comprehensively tangible due to their complexity. Unpredictable disturbance and inaccuracies lead to dynamic input conditions, which have an equally dynamic effect on the result.

Especially in the simulation, one tries to achieve reproducible results. This is in contradiction to the fact that the goal is to reproduce stochastic scenarios. This is only possible if disturbance effects are neglected or combined in the form of assumptions and simplifications so that all boundary conditions are always identical.

The term artificial intelligence is popularly used in an inflationary way and meanwhile conveys the image that a process can be carried out by machine-learning and is therefore subject to a strong blurring. Even simple algorithms of mathematics, which were previously referred to as such, now enjoy an upgrading through the categorization of AI, which is correct as such, insofar as they follow certain basic principles of intelligent functions. Expectations

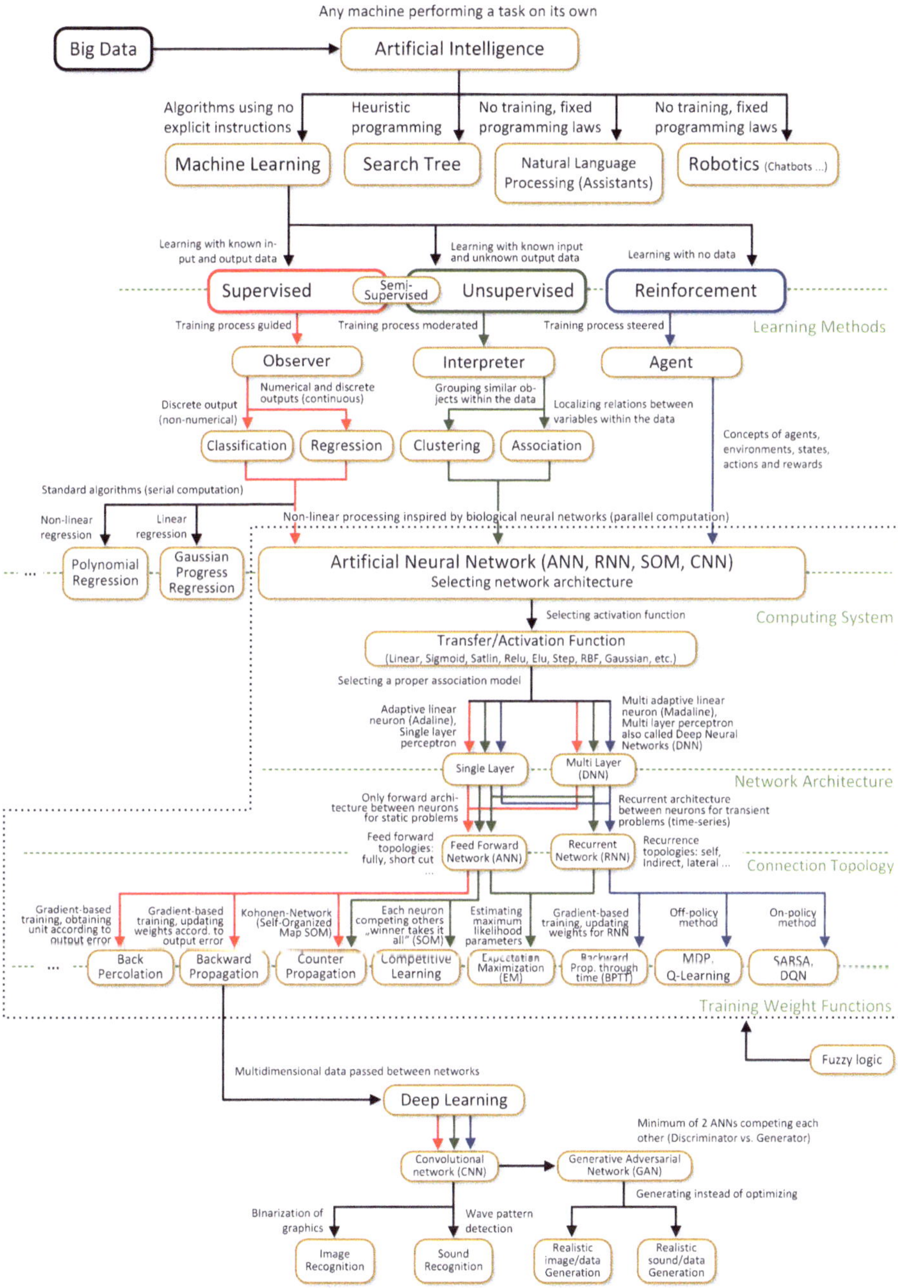

Fig. 5.5 Flowchart: Levels of artificial intelligence

of an AI model can vary greatly. Therefore, it is relevant to understand that we currently have various AI levels available, with varying degrees of model creation, individual degrees of complexity, and suitability for very extensive application fields. Looking ahead to this chapter, the flowchart shown above provides a detailed overview of the different levels of AI (Fig. 5.5). The chart also serves as a guideline, on which later contents in this book will continuously be referring to. Along the flowchart, all levels of AI are discussed in their basic features and general concept examples are presented, which are intended both to concretize the techniques and support the reader in stimulating his or her own creativity and to provide ideas for the development of his or her own application ideas.

5.3 Search Tree

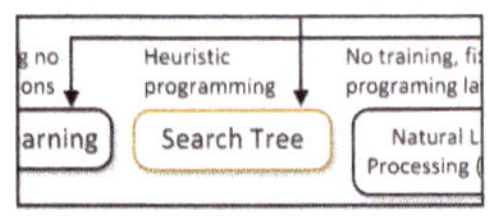

In the field of computer science, search tree functions belong to a long-established and important method of artificial intelligence. In an optimization process, search trees enable to transfer associative structures into an algorithm based on human decision-making. Among the most important are the A* (A-star) algorithms (Fig. 5.6).

A* algorithms are counted among the class of "informed search algorithms". The search for an optimum is formulated here in the form of weighted diagrams. From a starting node, the requirement is to find a path out of many possible ones that represent the lowest cost

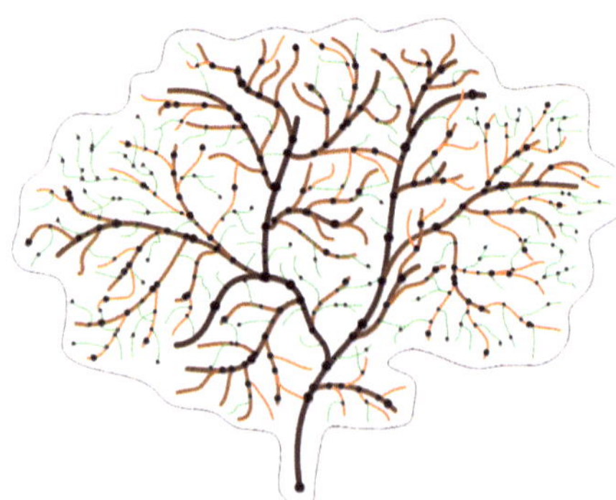

Fig. 5.6 Search tree structure

function. Cost can be any target quantity such as distance, time, efficiency, consumption and battery state of charge (SOC). If the starting node of the search tree is defined, the path of least resistance (here cost function) is extended until a termination criterion is met. The selection of the correct path together with possible and innumerable alternative paths is done by a heuristic method. In this context, the teaching of heuristics is to use limited knowledge to draw conclusions about a system by an analytical estimation procedure.

At every iteration of a main loop, A* has to define which of its paths shall be extended. This is based on the cost of the respective path and an estimate of the necessary cost to extend the path to the target. In particular, A* chooses the path that minimizes $f(n)$ with

$$f(n) = g(n) + h(n) \tag{5.1}$$

Here, the variable n denotes the next node on the path, $g(n)$ the cost of the path from the start node to n and $h(n)$ a heuristic function which estimates the cost of the most favorable path from n to the destination [41].

The search depth is limited by the available computing time. On the one hand, this is proportional to the computational effort of the evaluation function and, on the other hand, it grows exponentially to the size of the search tree. Fig. 5.7 shows the schematic structure of a heuristic downward strategy (HDS). Starting from the right, the heuristic search performs as many investigations until it has optimized a subject into the fifth level. The idea with an HDS is not to provide the same amount of computing time to each subsequent node in the path but to investigate promising paths more intensively. For this purpose, all possible successor nodes x are sorted according to their neural evaluation $V_N(x)$. The evaluation function can take many different forms, which will not be explained in detail here.

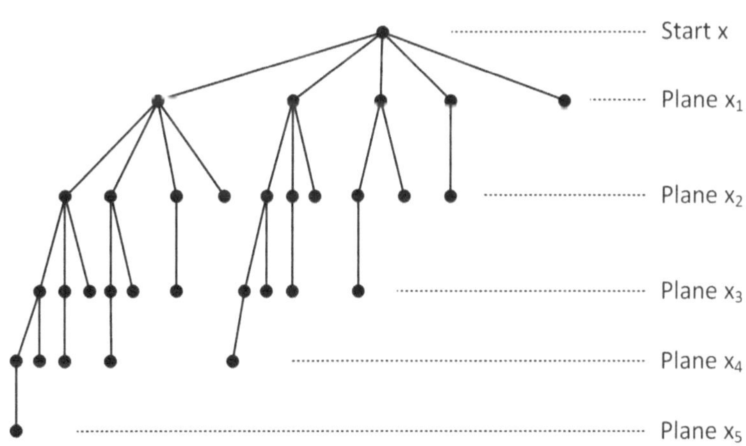

Fig. 5.7 Search tree through heuristic downward strategy [42]

$$HDS^1(x) = x_1$$
$$HDS^k(x) = f(HDS^{k-1}(x_1), HDS^{k-2}(x_2), \cdots, HDS^2(x_{k-2}), HDS^1(x_{k-1})) \quad (5.2)$$

The parameters $(x_1, x_2, x_3, \cdots, x_k)$ correspond to the successors of a node x in sorted order.

Similar to the way GPS systems work, such as those on the basis of updated driving data (weather station, traffic data) according to the heuristic search tree principle, multidimensional search trees can be nested together. Here, the result, i.e. the last node of a search tree serves as the start node of a new search tree. In GPS systems, this method of operation makes sense, since the consideration of all data over a desired route can become extremely large and the system would require a high computing time due to a massive amount of processes. The division into many different single stages can significantly reduce the workload, especially due to the continuous test procedure. Besides navigation systems, where search trees are used to determine the shortest route, they are also used in search engines. Their goal is to suggest the highest possible hits from a database when the user enters a search term. Further applications can be found, for example, in digital dictionaries, in the creation of complex database structures, for the prioritization of waiting loops or for compression algorithms of data formats such as JPEG and MP3.

The following are conceptual examples of how the search tree method can be used in the field of powertrain development (Fig. 5.8).

Example 1: Stationary application lean concept motor

This example shows how an ECU application can be performed on a lean concept motor using a search tree method. An optimization of parameters here is carried out on a stationary basis, i.e. after each working cycle on a warmed-up motor. There are a total of seven parameters objected to optimization: 1. air-fuel ratio, 2. valve timing on the intake side, 3. valve timing on the exhaust side, 4. wastegate position, 5. tumble position, 6. throttle position and 7. ignition timing.

The optimization by the search tree is done by this order and can go through several iterations until the result converges. The operating point will be predefined by a constant engine speed—optionally at controlled boost pressure through the wastegate position, which takes place in step 4 and/or additionally at controlled load, which takes place over the throttle position in step 6.

Each parameter is subdivided into 4 categories. The search tree gradually selects the optimal efficiency solution for each parameter variation and passes the best result to the next calculation step. In the first iteration loop x=1, the efficiency grows exponentially due to initially unfavorably selected starting conditions. In the second iteration loop, however, a fine-tuning process takes place, which reveals further potentials.

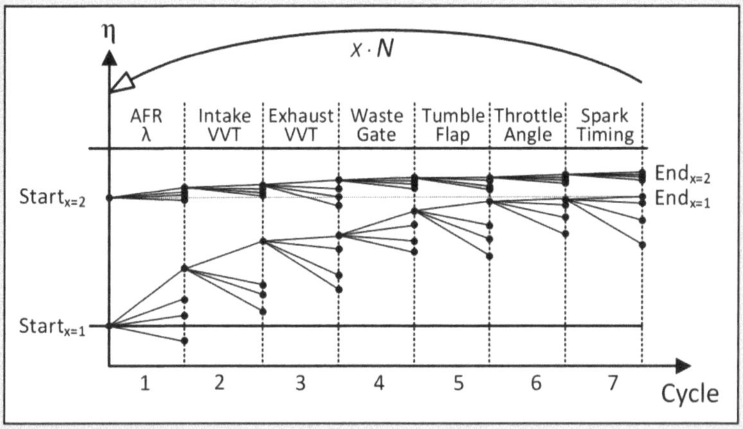

Fig. 5.8 Optimizing ECU application parameters for a lean concept engine using search tree

By anchoring the parameter sequence, this optimization method can also lead to instabilities, so that fluctuations prevent convergence of the calculation. In this case, it may be advisable to change the order of the parameters or to increase the number of categories to allow a finer approach to an optimal solution.[1]

Example 2: Operation strategy hybrid engine: Emission reduction in the warm-up phase versus loss of efficiency

The next example shows how the method of the search tree and the A algorithms can be used for transient warm-up optimization on a hybrid engine. Usually, low fuel consumption and low exhaust emissions are contradictory. To realize both at the same time is not possible, because the reduction of one causes the deterioration of the other.*

In a warm-up, for example, after starting an engine, it is import ant to warm up the exhaust system quickly by means of high exhaust gas temperatures so that the catalyst reaches its operating temperature (catalyst light-off), which is typically in the range $200°C$–$300°C$*. Only then does it start to convert pollutants such as* CO, HC *and* NO_x *efficiently. An internal measure that typically implements this is late ignition timing. When the exhaust valves are opened, an enthalpy flux at high temperature (unconverted energy to the crankshaft) is transferred. The exhaust gas heats the catalytic converter, and as a consequence, the engine efficiency deteriorates drastically.*

[1] The smaller the step size of the parameters chosen is, the more precisely not only local minima or maxima can be localized but also the interactions between the individual parameters can be decoded.

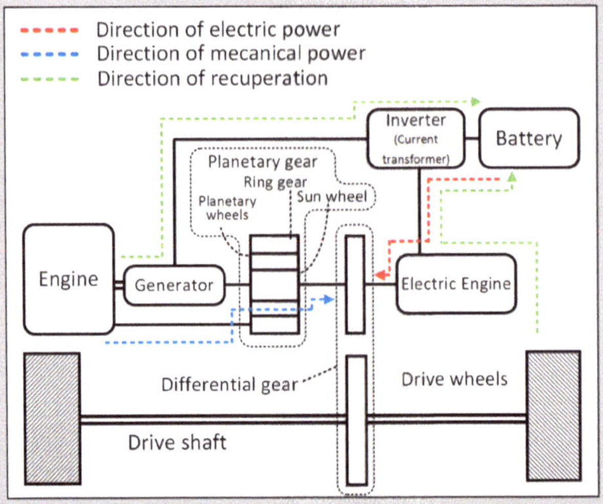

Fig. 5.9 Systemic representation of a parallel-serial hybrid powertrain

It is precisely here that the transient application of the search tree can provide efficient insights and suggestions as to how the hybrid strategy, i.e. the interaction between the combustion engine and electric motor(s) in a driving cycle, can be designed to find the best possible relationship between consumption and emissions. Fig. 5.9 *shows the schematic structure of a parallel-serial hybrid drive* [29].

The electrical path can be switched on or off through the parallel connection. The possibility of decoupling the electric motor results in a purely motorized, a purely electric or a mixed drive, which brings its individual advantages depending on the driving condition. The different torque sources can be added together by different types of transmission. In this example, not only a planetary gear is shown but also a direct coupling by means of spur gear or drive chain or by means of traction addition is conceivable. The battery can be charged either by increasing the load of the combustion engine (conversion of mechanical energy into electrical energy by the generator) or by recuperation, i.e. by using the braking energy. In order to use the braking energy, the electric motor takes on the function of a generator and feeds energy through the conversion chain from mechanical to electrical and finally to chemical energy and returns it to the battery.

In an experiment, for the first 200 s of the WLTC driving cycle, four different application strategies are provided to select from a search tree, which can alternate at each discrete time signal. The first strategy runs in an efficiency mode with an early ignition timing (SA) at knock limit, a closed wastegate in the acceleration phase and full support by the electric motor. The second strategy provides two intermediate solutions for all three parameters. The third strategy operates in a minimum emission mode, where the ignition timing is late, the wastegate position open and low electrical support to warm up the exhaust system quickly. The fourth operates in a recuperation mode, which starts charging the battery in braking and idling phases. During the transient run, the following integral is calculated using the sum of the effective engine efficiency and the catalyst efficiency.

$$\eta_{ges} = \int_{t_0}^{t} \frac{1}{2}\left(\eta_{eff} + \eta_{cat}\right) \tag{5.3}$$

The effective efficiency of the combustion engine is determined by the effective power P_{eff}*, the fuel mass flow* $\dot{m}_B$ *and the lower fuel heating value* H_u*:*

$$\eta_{eff} = \frac{P_{eff}}{(\dot{m}_B H_u)} \tag{5.4}$$

and the conversion rate of the catalyst from concentrations of the exhaust gas species at the inlet $K_{i,in}$ *and at the outlet* $K_{i,out}$ *by*

$$\eta_{cat} = \frac{\sum_{i=1} K_{i,out}}{\sum_{i=1} K_{i,in}} \tag{5.5}$$

According to the heuristic downward strategy, all four application strategies are executed in Fig. 5.10 at discrete sampling rates. After each new search tree, the variant with the highest cumulative efficiency η_{Ges} *is selected, from where the next search tree is performed. By means of this, the strategy with the highest efficiency can be determined along the complete search path with a low computational load. It presents the best compromise between engine efficiency and fast heating of the exhaust system to reduce emissions.*[2]

[2] The sampling rate can be selected more finely and more diverse parameters can be selected for the application strategies to ensure a more detailed breakdown for an evaluation than is shown here as an example.

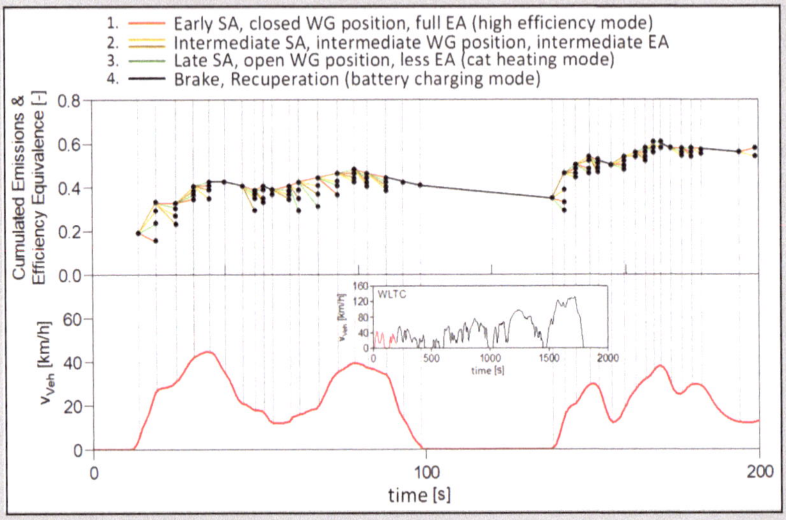

Fig. 5.10 Transient application of search tree: Optimization of a hybrid control strategy for the WLTC

5.4 Machine Learning

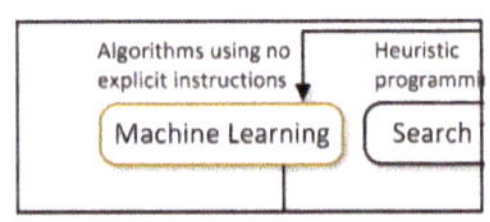

Machine Learning is a subarea of artificial intelligence. In contrast to explicit programming, it aims to enable systems to learn regularities based on data. The focus is on independent learning and the automated generation of a program code. In order to apply Machine Learning, large amounts of data are required. In this respect, the era of digitization and the resulting storage capacity and availability of enormous amounts of data (Big Data) is the actual driving force behind the development of Machine learning (Fig. 5.11).

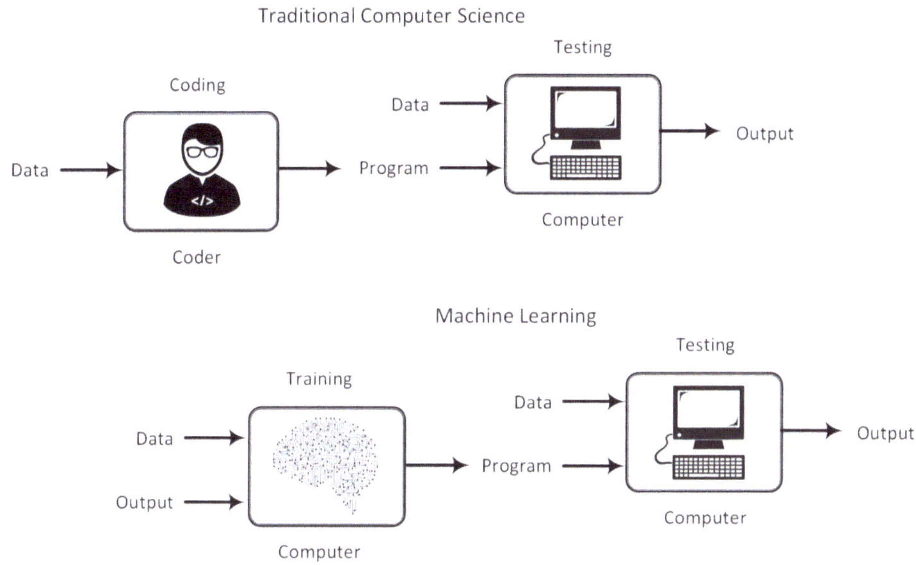

Fig. 5.11 Traditional programming versus machine learning

Machine learning includes many types of algorithms and training methods, all of which aim to generate statistical models that are able to analyze unknown data sets and make predictions based on their training data. Machine learning can basically be divided into three main learning categories: **Supervised Learning**, **Unsupervised Learning** and **Reinforcement Learning**, see Fig. 5.12 [43].

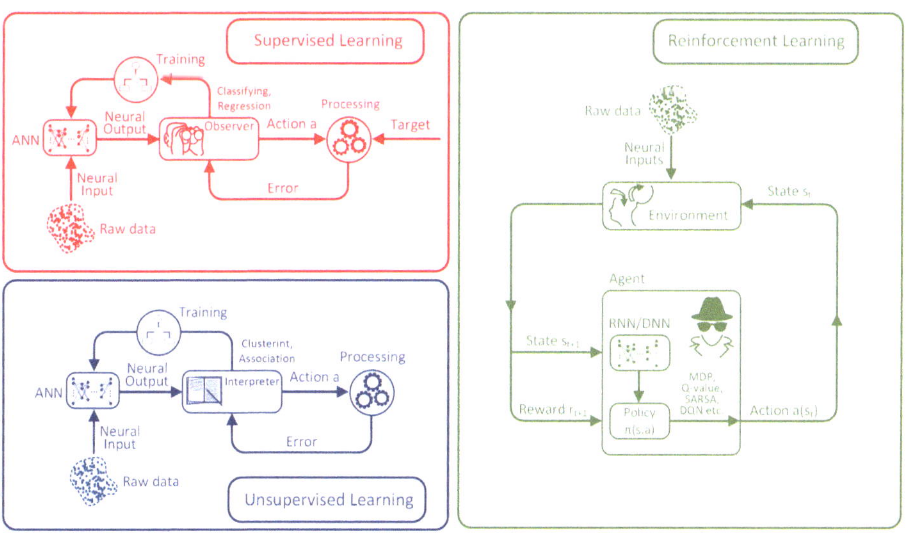

Fig. 5.12 Learning methods for machine learning

Only through the combination of machine learning algorithms and the system architecture of neural networks (see Sect. 5.5), both components combine to form a powerful overall system, which is today understood under the term artificial intelligence.

5.4.1 Supervised Learning

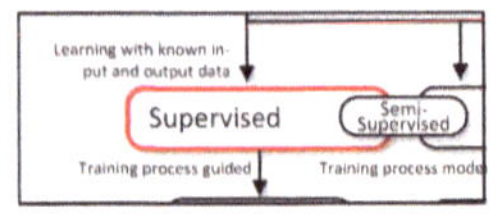

Supervised learning (SL) is one of the machine learning methods that requires a high level of interaction from the user during the training phase. For this method, both input data and output data in the form of target values must be known. Learning starts, as usual, with a set of training data selected from a larger data set. Part of the data is withheld from the training phase so that it can be used for a later validation phase. The learning method tries to recognize patterns between input variables and target variables within the data in order to transfer them into an analytical context. This can be done on the basis of different functions, polynomial approximation, Gaussian processes, or similar methods typically used in the field of powertrain development. These methods have proven to be particularly useful when real-time data generation is desired. Neural networks form a completely new basis of a model architecture, which can be used to design application paths with exponentially higher performance and process speed.

Classification and Regression

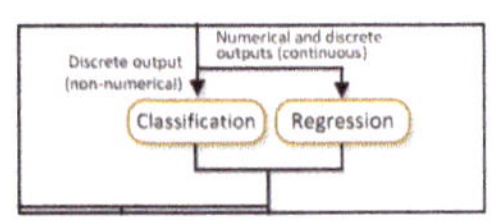

In order to analyze the significance of the data fed in, properties are first assigned to them. If data is discrete and can be categorized, a so-called **classification** is a suitable method in terms of the supervised learning method. If, on the other hand, a data set is numerical and continuous, the solution can be found by a **regression**. Fig. 5.13 illustrates both application types of supervised learning.

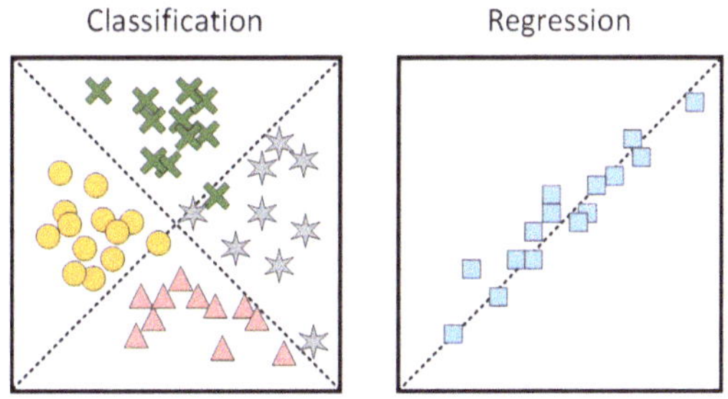

Fig. 5.13 Classification and regression of data for the supervised learning process

Observer

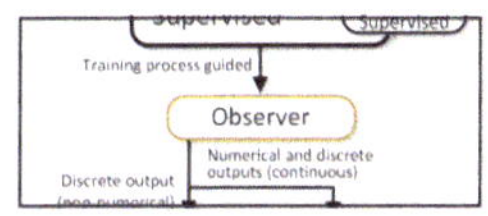

Machine learning focuses on the ability not only to reproduce results but also to steadily improve a solution path. The task of a learning system is to develop strategies to optimize behavior during the training phase. Within the framework of the supervised learning procedure, a so-called observer is used for this purpose. This observer is assigned the task to constantly monitor the error between the output of a model and the target. A simple form of strategic learning is to minimize an error between the target and the actual value. Fig. 5.14 shows schematically how the logic of such an SL procedure can be constructed in combination with an artificial neural network (ANN).

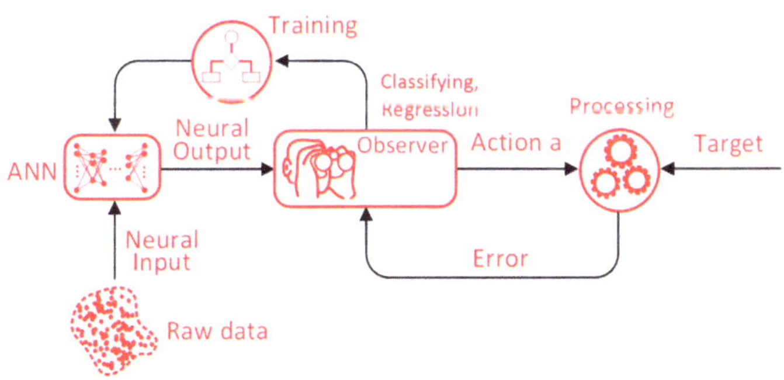

Fig. 5.14 Observer for the supervised learning process

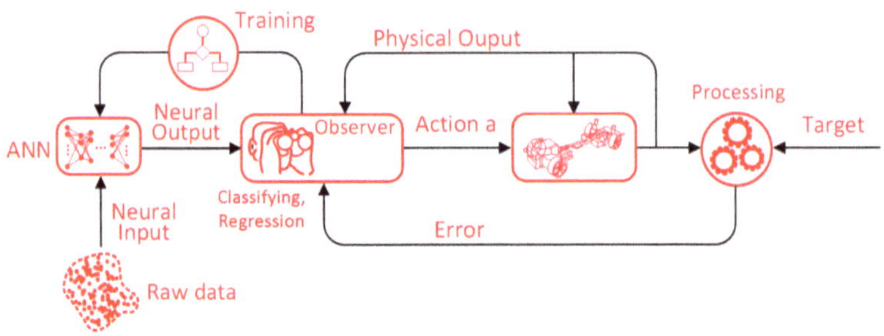

Fig. 5.15 Observer combined with a physical model

Often, it is desired to anchor parts of a calculation process in neural networks while retaining others on a physical level. This hybrid approach can be exercised by using the supervised learning method. The interaction of both systems proves to be very powerful, especially under the aspect that a physical mapping layer is transparent and comprehensible, whereas the ANN layer follows mathematical principles, and its decision-making is fundamentally unexplainable (Fig. 5.15).

Currently, the supervised learning method is widely used and is applied in economical sectors, for fraud detection, risk analysis for banking and financial sectors, product and job recommendation advertisements, Internet commerce, spam detection, speech recognition and many more.

The following examples are presented for possible applications within the powertrain development, where supervised learning is well implementable.

Example 3: Engine flow, combustion (classification)
An application from the field of fluid dynamics is the determination of the characteristics of flows. Based on physical evaluation variables, these can be classified either as laminar or turbulent or in possible subcategories. This is relevant, for example, to a) calculate heat transfer in the air path of an engine in order to draw conclusions about heat losses and thermal behavior in general, or b) to determine the transition from laminar to turbulent flow for fluids. The critical Reynolds number, which marks the transition between laminar and turbulent flow, depends not only on the geometry of the application case (tube inner flow, tube outer flow, plate flow, etc.) but also on the

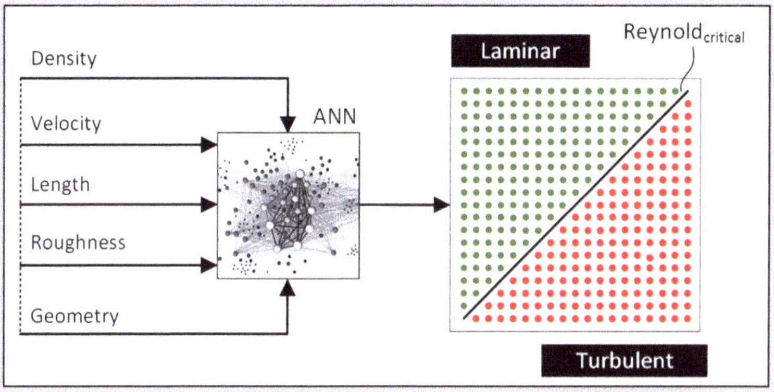

Fig. 5.16 Classification of thermokinetic data (Reynolds number)

choice of the characteristic length. The decision about laminar or turbulent or about subcategories (0 < Reynolds number < 200, 201< Reynolds < 400, ...) is here delimited by threshold values in classes and is therefore suitable for a classification from the category supervised learning (Fig. 5.16).

Example 4: Thermokinetics (Regression)

A large part of all simulation tasks in the area of powertrain development deals with regression problems. Especially when calibrating models, the first step is often to map real data using models with a high prediction quality. In the following, a typical application is presented on how the combustion engine can be used as a combination of several submodels in an ECU system. The engine in Fig. 5.17 consists of the submodels "fuel", "air path", "torque" and " emission". Since a crank angle-resolved calculation of the variables is not necessary, the results can be generated faster than in real time.

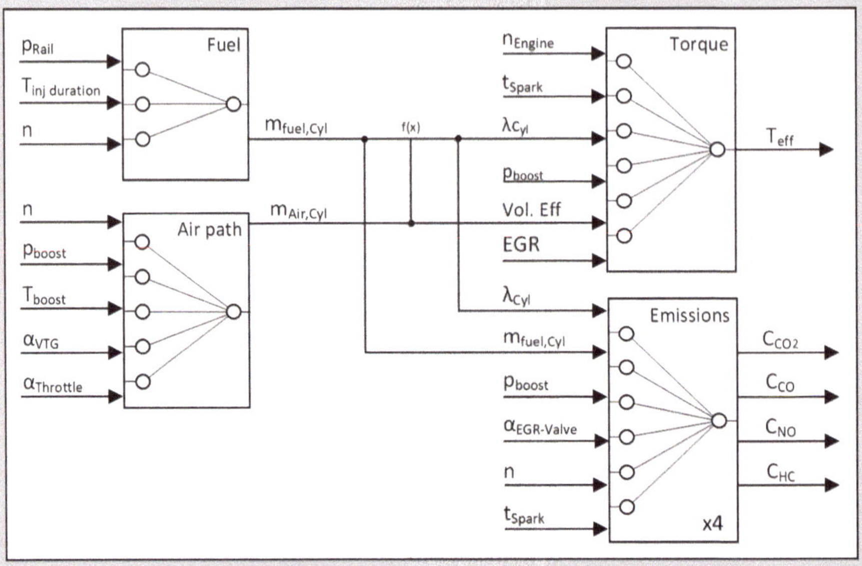

Fig. 5.17 Engine model of a control unit [44]

Fig. 5.18 presents the result of a regression analysis for the torque module (top right) based on 600 data points. The torque is shown as a function of the input parameters speed (n_{Engine}), spark timing (t_{spark}), air-fuel ratio (λ), intake manifold pressure (p_{boost}), air volumetric efficiency (Vol. Eff) and exhaust gas recirculation rate (EGR).

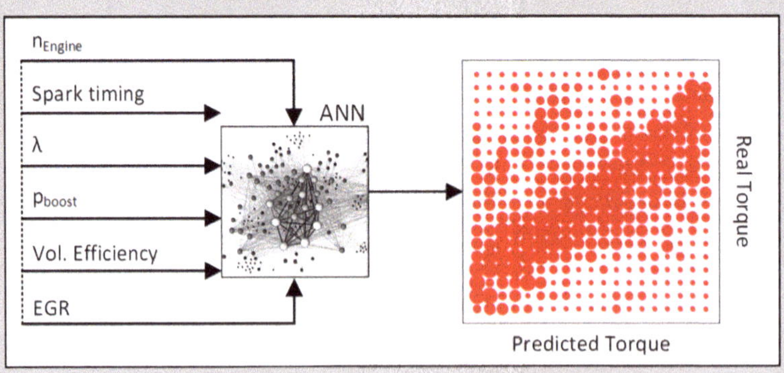

Fig. 5.18 Regression of thermokinetic data (Reynolds number)

5.4.2 Unsupervised Learning

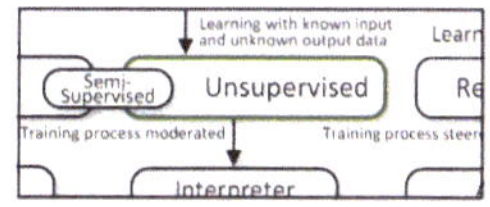

Unsupervised learning is a powerful method that is suitable when the amount of data to be examined is considerably large and, above all, not clearly identified. In principle, it can be composed of both discrete and continuous entries. The learning procedure is useful if the user lacks the coherence of data, if it is difficult to label the data or if the data does not contain any output variables. In contrast to the supervised learning procedure, the focus is not on forming a correlation between input and output data, but on characterizing features within the data. The strength of the learning procedure is based on the decoding of hidden patterns. The type of decryption is divided into two categories: The "clustering" and the "association".

Clustering and Association

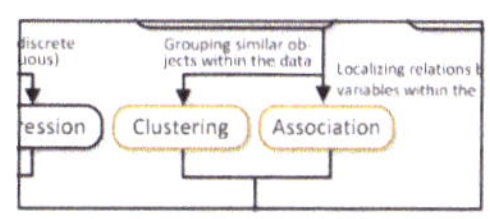

Clustering is, as the word implies, the grouping of objects that show similarities in their characteristics. The so-called associative rule learning goes one level deeper and localizes connections between existing parameters within the database. Unsupervised learning performs an iterative and autonomous analyzation process without any user intervention. Fig. 5.19 illustrates the processes of clustering and association.

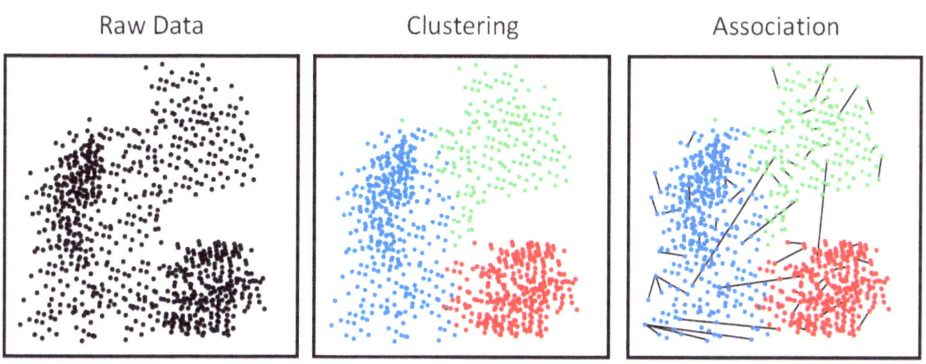

Fig. 5.19 Clustering and association of data for the unsupervised learning procedure

Interpreter

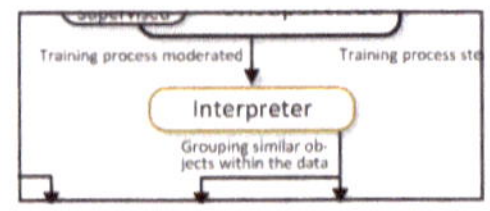

For the unsupervised learning procedure, an interpreter is generally used to process the Clustering and Association. The following diagram illustrates how the training loop is connected for such a procedure (Fig. 5.20, 5.21 and 5.22).

Ideally, an unsupervised learning process can also be used as a preliminary stage to mark unknown and unclassified data and then pass it on to a supervised learning algorithm (semi-supervised learning). Typical use cases for unsupervised learning can be found in the analysis and segmentation of customer behavior, market data, and sales data and so to optimize market activities. In healthcare medicine, the constantly growing volume of patient data is used to correlate disease findings with individual complaints and symptoms, resulting in training more reliable models that can be used to detect diseases at an early stage and provide timely treatment.

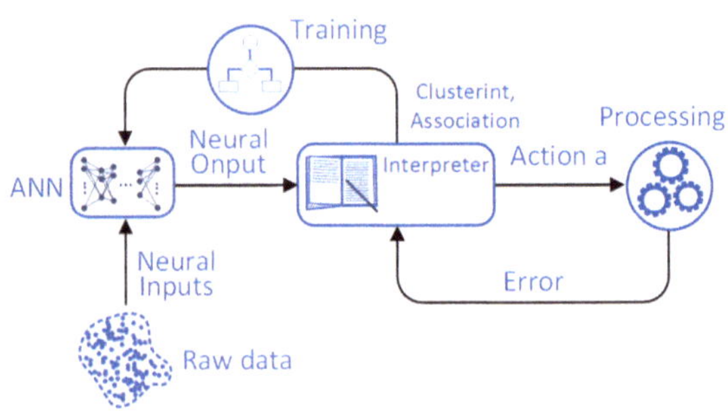

Fig. 5.20 Interpreter for the unsupervised learning process

Example 5: Pre-processing of measurement data/real data (clustering)

The handling of large amounts of data is one of the core issues that poses great challenges for engineers in many phases of powertrain development. In retrospect, as shown in Fig. 4.13, a model-based development process consists of recurring loops in which measurement data must be analyzed and post-processed. The process of selecting usable from unusable data can be very time-consuming and is often subject to the user's decision criteria. The clustering method of the unsupervised learning procedure proves to be extremely powerful for such applications. It is capable of grouping unmarked data and reliably identifying measurement outliers by a standardized process.

In the first step, data analysis can be carried out after inputting measurement data (here using the example of 10,000 data points and 10 variables each). The quantization error (QE) with the value range [0,1] describes the error between a data input and the value that a neuron outputs as a result after training. According to the error, the data is first broken down along a normal distribution. Using a threshold value, the user defines the aggressiveness with which the data selection should be performed. If a threshold with the value x is defined, the corresponding proportion of $(1 - x)$ *data is sorted out of the data set (Fig. 5.21).*

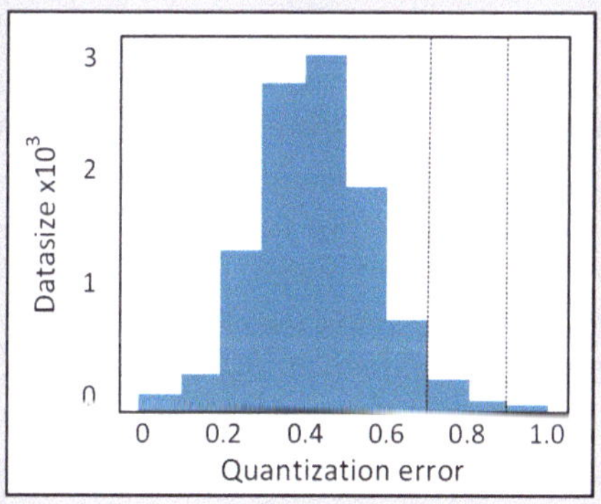

Fig. 5.21 Normal distribution of measurement outliers (quantization error) of a data set

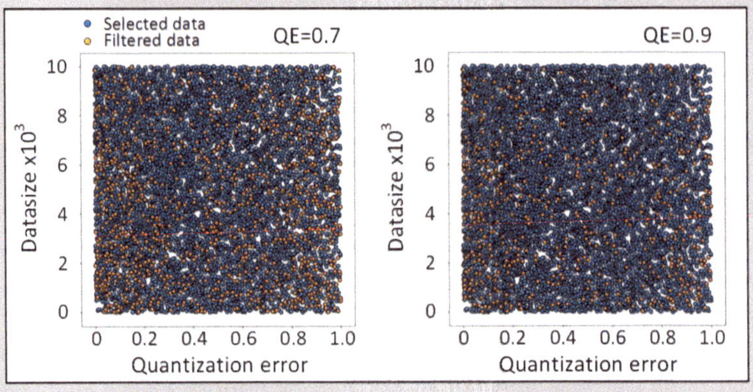

Fig. 5.22 Data selection for two different quantization error thresholds

Using the example of two selected threshold values (QE=0.7 and QE=0.9), the result of the data selection is given in Fig. 5.22.

Example 6: Interpretation of parameter studies—fuel cell efficiency (clustering)

The focus of data analysis is always the determination of influencing factors of specific input variables on one or more target variables. If there are many input variables that show a nonlinear correlation to a target variable and if these variables show additional cross effects among each other, the system is highly complex and cannot be easily interpreted without additional monitoring algorithms. Unsupervised learning can provide an enormous remedy for such cases. It is able to project data features of different kinds and especially their combinations to target values. If such a model is designed for a specific situation, it can provide fast and reliable statements regarding the data features in the future by entering new data sets.

Using the example of a fuel cell, a parameter study is carried out on the basis of a measurement series of 50,000 data sets. For this purpose 9 parameters, which determine the efficiency, are combined with a Latin hypercube DOE method (see Fig. 4.27). These are 1. temperature, 2. air mass flow, 3. air pressure, 4. humidity, 5. cathode resistance, 6. anode resistance, 7. capacitor capacity, 8. cathode stoichiometry and 9. anode stoichiometry. A self-organizing map (SOM) is trained using the Competitive Learning Procedure (see Sect. 5.5.3). As a solution, this map categorizes all possible combinations of inputs into three efficiency classes: high efficiency (1), intermediate efficiency (2) and low efficiency (3) (Fig. 5.23).

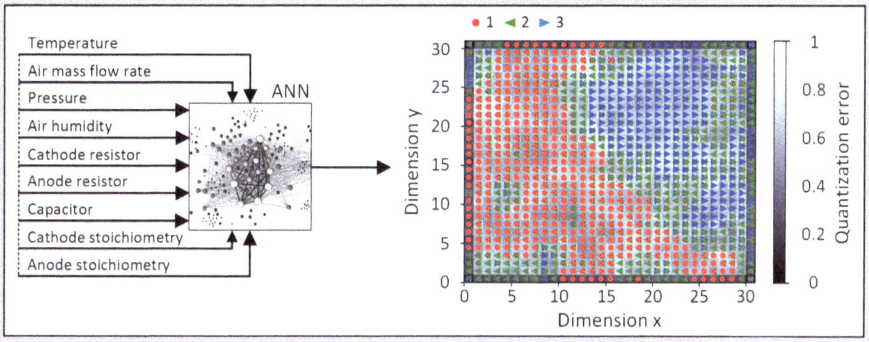

Fig. 5.23 Fuel cell: Grouping of efficiency classes

Example 7: Material Load Analysis Method (Clustering)

Another possible application of clustering is for failure analysis in the field of materials science. By entering a load type of a component (static or dynamic), the load time until material failure (fracture or deformation) and the fracture type (deformation-free, low-deformation or high-deformation) or the deformation type (plastic or elastic) and the characteristics of materials can be transferred into a data matrix. Clustering enables the introduction of an analysis procedure that allows component failures to be examined in more detail. For example, when screening a damaged component, clustering can tell which material with which specific alloys (Fe, Cr, Mo, Ni, Si, C, etc.) the component consists of (Fig. 5.24).

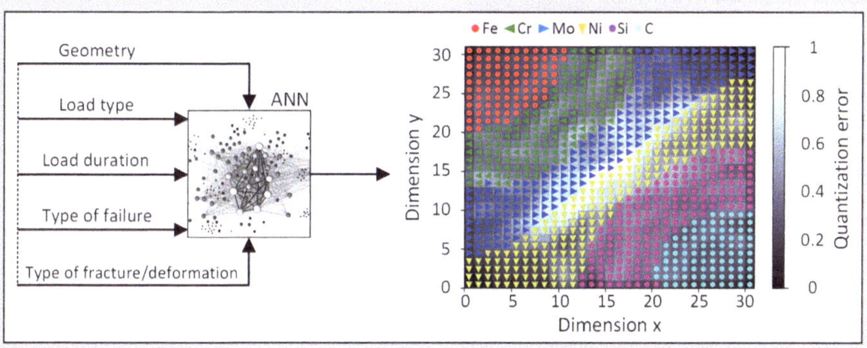

Fig. 5.24 Clustering to investigate a steel with nickel-chromium-molybdenum alloy after material failure

5.4.3 Reinforcement Learning

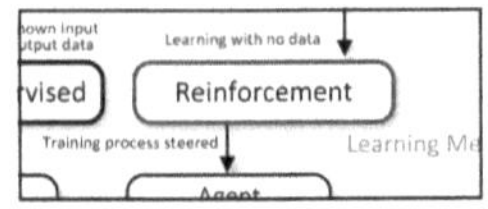

Reinforcement Learning (RL) is a further learning procedure next to supervised and unsupervised learning that is particularly characterized by its high performance and self-learning properties. RL uses a so-called "agent" as its main feature, which actively interacts with its environment, which has a set of states, independently searches for strategies (policies) and executes actions to maximize the accuracy of a model and is rewarded as a result of correct learning. While the supervised learning and unsupervised learning methods rely on existing real data during the training phase, the RL is highly dynamic and adapts to constantly changing data inputs. This is achieved through the interaction of the agent with its environment, which results in a continuous strategy adjustment and efficiency increase. Due to its self-learning properties, the system requires little to no real data and can still achieve resilient results.

When talking about artificial intelligence, users often have the idea that a smart algorithm is able to achieve the same solution faster and more efficiently than humans already achieve. If the user has a clear idea of what a result should look like, this leads to a strong limitation of learning. The challenge for AI would then be to find suitable ways to replicate the same solution. If, on the other hand, there are no concrete expected variables in a given situation that should be correlated by means of input variables, an additional degree of freedom results for the algorithm, with which solutions can be explored that go beyond the human imagination and which the user does not expect. Reinforcement learning is particularly suitable for this type of case.

Fig. 5.25 Agent for the reinforcement learning process

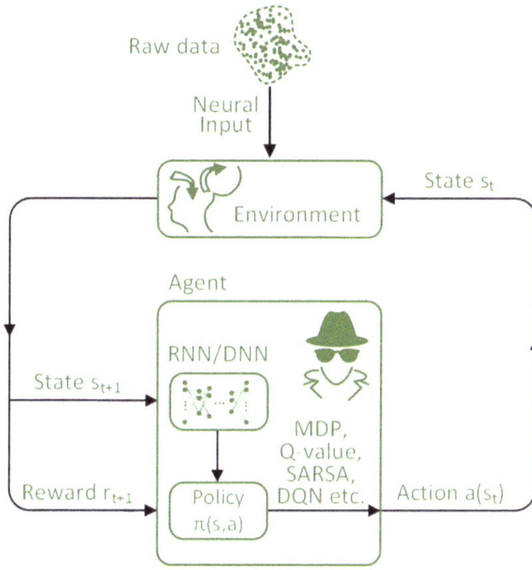

Agent

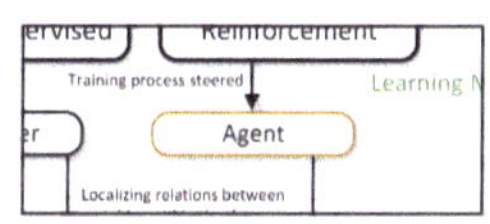

During data analysis, the algorithm receives feedback after each iteration loop from an agent, which here takes the function of a teacher. In contrast to the supervised or unsupervised learning methods, the algorithm does not only learn from the data analysis but also from an additional trial-and error method, which gradually reinforces a mapping function for predicting a result. This has the advantage that a solution to a problem can be found via a smaller number of iteration loops (Fig. 5.25).

Example 8: Adaptive speed controller

Since September 1, 2019, the real driving emission (RDE) cycle is the certification basis for all new vehicles. Along with the worldwide harmonized light vehicles test procedure (WLTP), this should lead to more realistic consumption and emission parameters. Unlike previous certification cycles, the RDE cycle will not reflect an average real-life journey. Both the route and the load as well as driving style are flexible. This means that the route, speed and elevation profile are not fixed and can be set individually by any inspector.

One of the many fields for which reinforcement learning is suitable is the adaptive design of controllers with static control parameters so that they can always be dynamically adapted to the operating state for transient applications. Especially against the background of variable cycle profiles, adaptive control is particularly suitable for both model-based and test-based development of powertrains on the chassis dynamometer. This allows a wide variety of profiles to be run without the need for time-consuming and constantly recurring calibration of the control parameters. For a fast and reliable determination of corresponding consumption and emissions, this can result in a great time advantage.

The following figure shows a Fig. 5.26 simplified scheme of a virtual, static speed controller. In the first step, the desired driving speed v_{req} *is converted into a desired torque* T_{req}*, taking into account existing gear ratios and existing driving resistances*[3] *of the corresponding vehicle. With the previously calibrated control parameters of a PI controller (proportional component* K_p *and integral component* K_i*), the throttle is controlled by the accelerator pedal with the goal of minimizing the difference between* $(T_{req} - T_{act})$*.*

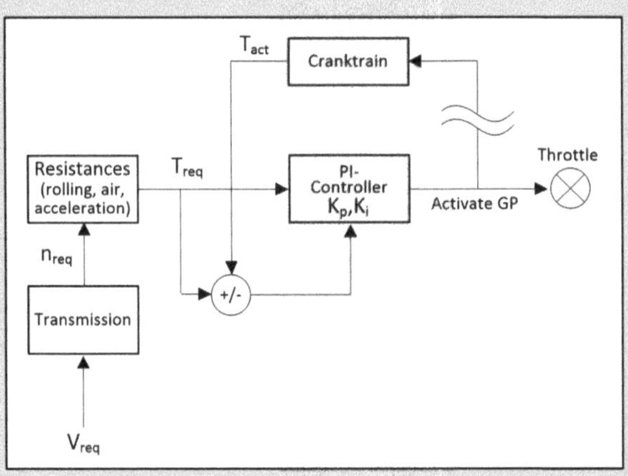

Fig. 5.26 Static control circuit for the vehicle speed of a cycle

[3] The driving resistance is usually divided into rolling resistance, air resistance, rotational acceleration resistance of all rotating parts and translational acceleration resistance of the total vehicle mass.

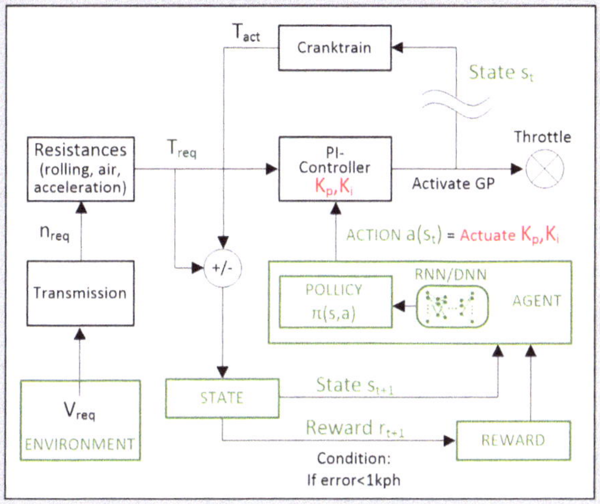

Fig. 5.27 Adaptive control loop through reinforcement learning

Since the control parameters are chosen to be constant, it must be assumed that even with good prior calibration, they will only work well in the range of the calibration. Due to the different operating conditions (speed and load) and the general variability of the RDE profile in testing mode, not all areas can be covered satisfactorily. If it is desired to use the reinforcement procedure, which automatically transfers an optimal set of parameters to the powertrain controller for each operating state, it is fundamental to define and implement the Agent, Policy, State, Reward and Environment functions into the control loop (Fig. 5.27).

In Fig. 5.28, the result of the above control loop is shown. The RL method needs a certain training time until it captures the control logic. From there, it is able to gradually adjust the given driving speed of the RDE profile with a small deviation (error $\leq$ 1 km/h). The ratio between the control parameters K_i / K_p is plotted below. For this purpose, a simple neural network architecture is selected. An increasing reward function describes the self-learning training effect.

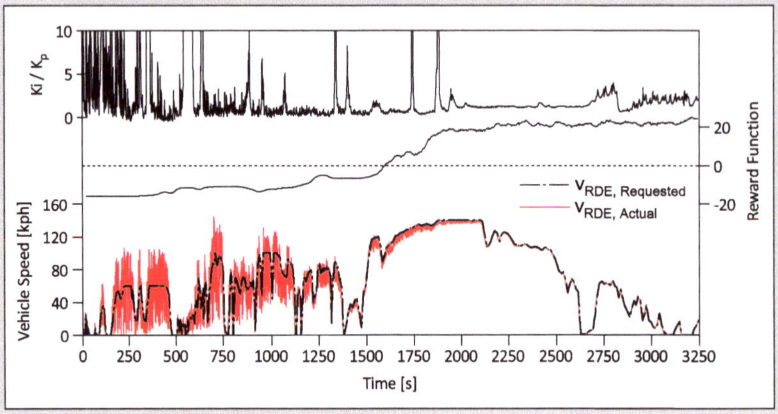

Fig. 5.28 Adaptive RDE speed control

Example 9: Vehicle chassis

A small company called Hackrod, based in California, is specialized in advanced chassis manufacturing for sports cars. The focus of their work is to free the limited design ideas of the human mind and to develop creative concepts based on digital training data and machine learning algorithms.

Using the RL process, the company succeeded in designing a chassis that is significantly more stable than a standard chassis, 30% lighter and therefore more cost-efficient. The net structure shown in Fig. 5.29 is asymmetrical, which is explained by the fact that the center of gravity of a racing vehicle including the driver does not necessarily have to lie on the axis of symmetry. The algorithm takes this into account and builds more heavily loaded regions with a correspondingly denser mesh than less-loaded ones.

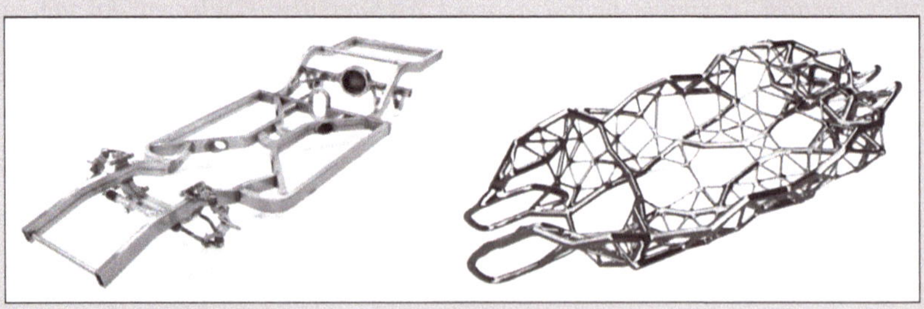

Fig. 5.29 Reinforcement learning applied to chassis design

5.5 Artificial Neural Networks (ANN)

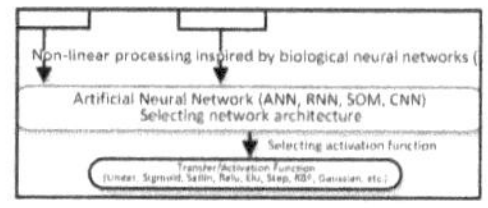

The elementary processing units of a brain consist of neurons. These are cells that exchange signals with each other through electrochemical impulses and thereby excite each other. The brain takes care of the reception, processing and response to stimuli, which are received by sensory organs and transmitted through signals. These signals are called receptors. They register all kinds of information such as light, temperature, sounds, thoughts, tastes and smells and transmit them to the brain via nerve tracts. There are about 100,000 neurons per square millimeter of the human cerebral cortex. One neuron has an average of 10,000 connections to neighboring neurons, so that a brain can have up to 10^{14} connections in total. The sum of these neurons forms a biological neural network (BNN) [45].

The processing and extraction of information in computer systems are different from that in a human brain. In 1945, the essential works of the mathematician John von Neumann were published, which formed the basic mode of operation of the computer architecture used until today. The so-called von Neumann architecture represents the individual components on which the mainboard of computers is built (Fig. 5.30).

While the calculation basis in computers is serially ordered, which leads to a lower data processing speed, processes in our brain are handled in parallel. Especially, the associative mode of operation of the brain allows creative and cross-topic approaches to solving problems, which is conditionally possible with a computer due to its address-based and 1D approach (Table 5.2).

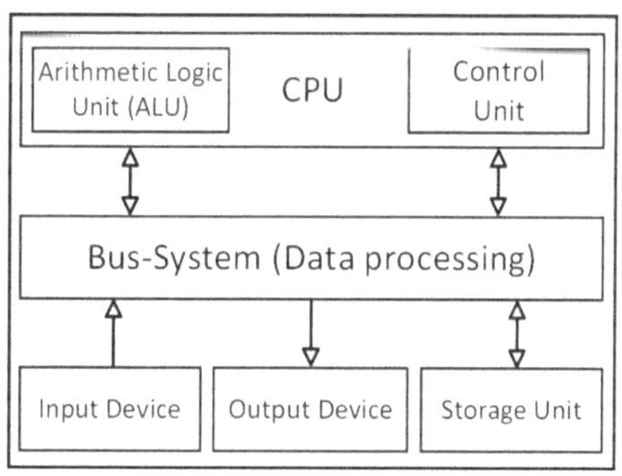

Fig. 5.30 Functioning of today's computer architectures based on Neumann [46]

Tab. 5.2 Data processing brain versus computer [47]

	Brain	Computer
Number of data processing units	$\sim 10^{11}$	$\sim 10^{9}$
Type of processing	Neurons	Transistors
Calculation configuration	Parallel	Serial
Data storage	Associative	Address based
Switching time	$10^{-3} s$	$10^{-9} s$
Number of possible toggle operations	$10^{13} \frac{1}{s}$	$10^{18} \frac{1}{s}$
Actual toggle operations	$10^{12} \frac{1}{s}$	$10^{10} \frac{1}{s}$

This results in many problem genres that cannot be transformed into an algorithmic form, or only with a great deal of effort, so that a computer can solve them. Unlike humans, computers do not experience a learning process and are not adaptive to recurring problems.

Based on human thought processes, a neuronal network of the cerebral cortex can be transformed into a model so that it becomes tangible through a mathematical formalism. Cell extensions of nerve cells (also called dendrites) emerge from a cell body, which serve to absorb stimuli and transmit them to neurons. Before they reach the neuron, they are inhibited or reinforced by synapses. The strength of the synapses can be described by so-called weights $w_{i,j}$, which can have any positive value $[0, \infty]$. A neural network basically consists of three layers, the input layer, the hidden layer and the output layer. Between the input values $I_1, I_2, \ldots, I_i$ and the output values $O_1, O_2, \ldots, O_o$, hidden layers can have any complex dimensions $L_1, L_2, \ldots, L_L$, which in turn can assume any number of neuron levels $N_{1,1}, N_{1,2}, \ldots, N_{1,N}$. The weights lie on the connecting line between two neurons, which is amplified or weakened by $w_{i,j}$, see Fig. 5.31.

A neuron is a processor that can assume a binary state. Either the state is deactivated or activated $I_{j,k} \in [0, 1]$. The product of the inputs $I_{j,k}$ with their respective weights $w_{i,j}$ forms a result that is measured at the threshold value Θ (bias). A neuron is then activated for following condition:

$$\sum I_{j,k} \cdot w_{j,k} > \Theta_k \tag{5.6}$$

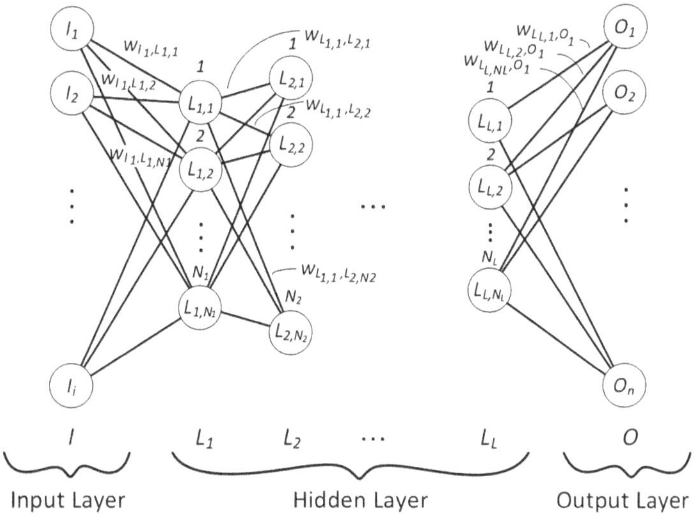

Fig. 5.31 Components of a neural network

Finally, a function s_k is introduced, which is the product of the input values $I_{j,k}$ and the weights $w_{j,k}$ plus the BIAS value:

$$s_k = \sum I_{j,k} \cdot w_{j,k} + \Theta_k \tag{5.7}$$

5.5.1 Activation Function

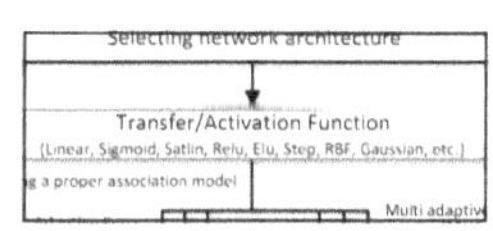

Activation functions are transfer functions that are used to calculate the weights. They are used to design nonlinear functional relations between input and output signals of neurons. An activation function is designated with F. After a signal s_k has been passed to a function F, z_k represents the result of the transfer:

$$z_k = F(s_k) \tag{5.8}$$

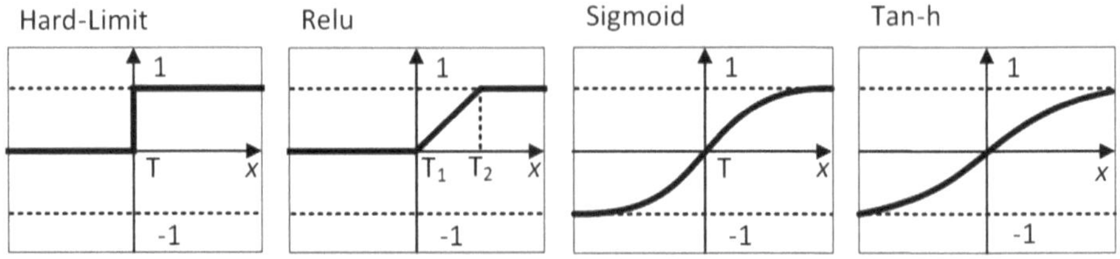

$$F(x) = \begin{cases} 1, & x > T \\ 0, & x < T \end{cases}$$

$$F(x) = \begin{cases} 0, & x < T_1 \\ \frac{x - T_1}{T_2 - T_1} & T_1 < x < T_2 \\ 1, & x > T_2 \end{cases}$$

$$F(x) = \frac{1}{1 + exp(-c(x + T))}$$

$$F(x) = \tanh(x)$$

Fig. 5.32 Activation functions

The choice of a suitable activation function for a neural network can strongly influence the quality of the model. The right choice is subjective and depends on what kind of data is used and what properties it has. Among the most prominent functions are Hard Limit, ReLU, Sigmoid and tanh. If, for example, an output value is to be generated in a binary form, Hard Limit is suitable. If the right term from Eq. 5.7 assumes a positive value, then s_k is given the value 1, otherwise $s_k = 0$ [48].

However, due to the step response at x = 0, the system becomes non-differentiable. This can cause system oscillations depending on the application. If the differentiability of the transfer function is desired, a ramp function rectified linear unit (ReLU), a sigmoid function or a tanh function (tangens hyperbolicus) can be used. Furthermore, there are averted activation functions such as Leaky ReLU, exponential linear unit (ELU) or scaled variant of ELU (SELU), which can bring individual advantages for particular application problems.

The architecture of a neural network is composed of many parallel running computational processes. Each unit in the network performs a small part of the overall process. It receives inputs from neighboring units or external sources, calculates a new output value and passes it on to other neighboring units. An overview of all internal processes between a set of inputs and a neuron is shown in Fig. 5.33.

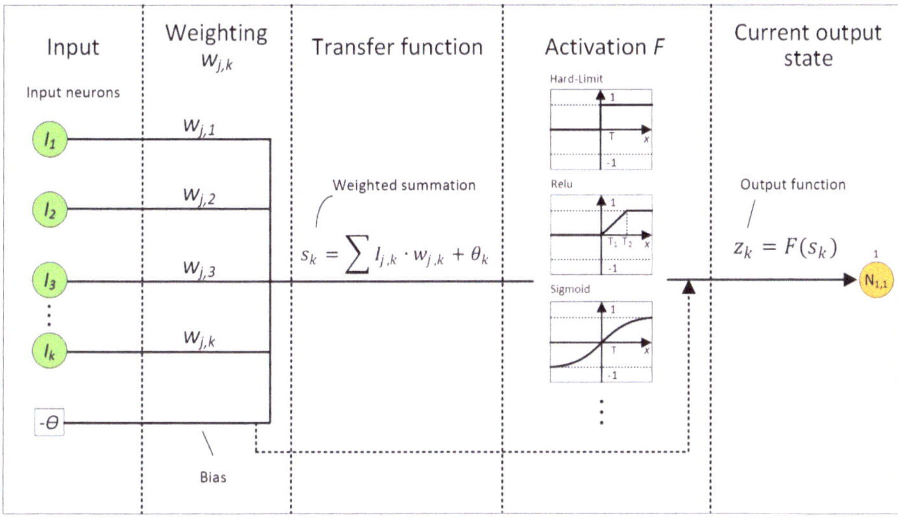

Fig. 5.33 Schematic representation of internal processes between two connected neurons [48]

5.5.2 Feedforward Networks (ANN) and Recurrent Networks (RNN)

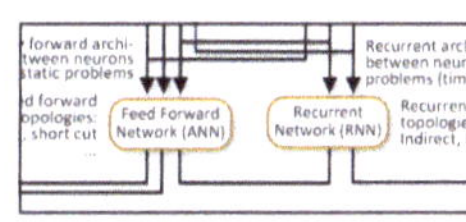

After the structure and function of a neural network were presented in Chap. 5.5, this chapter deals with possible network topologies. Topology means the variety of possible connection patterns between the individual neurons. A fundamental difference is made between a **feed-forward network** and a **recurrent network**.

In a feedforward network, as the name indicates, the flow of information points in a forward direction from input to output. Single-layer perceptrons belong to a subgroup of feed-forward networks. These include network structures that, unlike the one shown in Fig. 5.31, do not have hidden layers. The input layer is thus directly linked to the output layer. Multi-Layer Perceptrons on the other hand consist of at least one hidden layer. Since this allows realizing much more complex structures and interconnection logics between neurons, multi-layer perceptrons are one of the most common network types today.

To illustrate possible network topologies, Hinton diagrams are often used in this context, which show the connections between neurons in a matrix form. Fig. 5.34 (left) shows a simple three-level network with an input layer I_i, a hidden layer h_i and the output layer a_i. As shown here, one often encounters feedforward networks, where each neuron i is

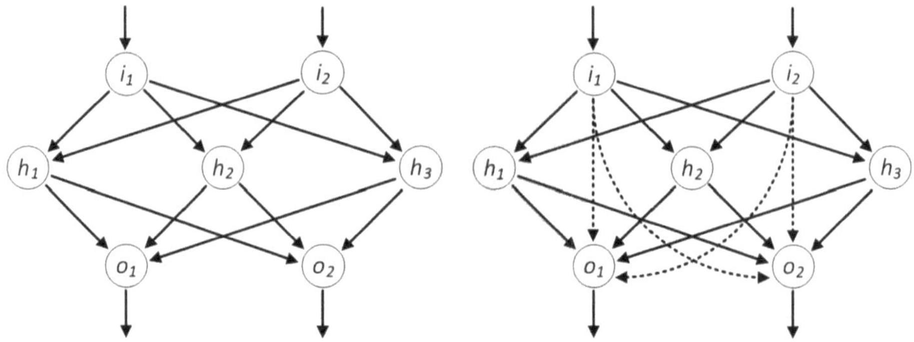

↱	i_1	i_2	h_1	h_2	h_3	o_1	o_2
i_1	0	0	1	1	1	0	0
i_2	0	0	1	1	1	0	0
h_1	0	0	0	0	0	1	1
h_2	0	0	0	0	0	1	1
h_3	0	0	0	0	0	1	1
o_1	0	0	0	0	0	0	0
o_2	0	0	0	0	0	0	0

↱	i_1	i_2	h_1	h_2	h_3	o_1	o_2
i_1	0	0	1	1	1	1	1
i_2	0	0	1	1	1	1	1
h_1	0	0	0	0	0	1	1
h_2	0	0	0	0	0	1	1
h_3	0	0	0	0	0	1	1
o_1	0	0	0	0	0	0	0
o_2	0	0	0	0	0	0	0

Fig. 5.34 Feedforward ANN with 3 levels: Hinton diagram with pure feedforward formation (fully linked network, left) and additional direct connections between an input and an output (short-cut connection, right) [47]

connected to all neurons of the next layer. This type of network is called a "Fully Linked Network" (left). Other feedforward networks allow so-called "short-cut connections", where connections can skip one or more levels (right).

In contrast to conventional feedforward networks, complex recurrent networks allow neurons to influence upstream and downstream levels and themselves through connections. This gives the system an iterative character, so that stability can only be achieved by certain iteration loops. The number of iteration loops depends on how well the starting conditions are chosen. The big advantage of RNNs is that they get a kind of memory by their iterative structure. This makes them particularly suitable for time-dependent data sequences (time series). Especially in the field of powertrain development, these powerful network structures allow solving transient problems.

In Fig. 5.35 (left), a network with a "Self-Recurrence" character is shown. This allows neurons to inhibit each other during the self-training phase until they are strong enough to exceed the threshold Θ and thus be activated.

If connections from a hidden layer are established back to the input layer, this is called "Indirect Recurrence". Unlike Direct Recurrence, here neurons can train their behavior as a consequence of downstream neural layers (see right graph).

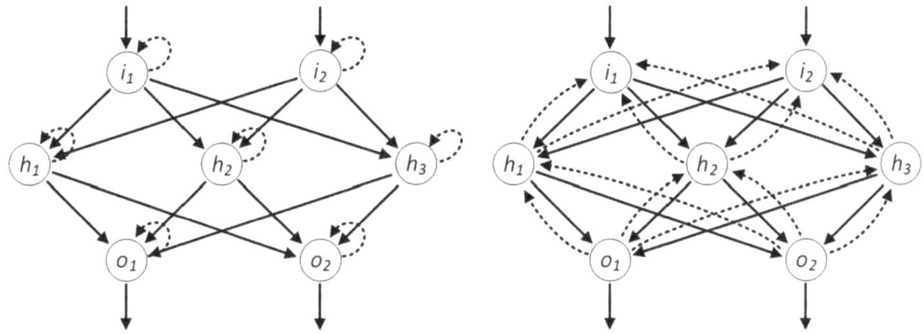

	i_1	i_2	h_1	h_2	h_3	o_1	o_2
i_1	1	0	1	1	1	0	0
i_2	0	1	1	1	1	0	0
h_1	0	0	1	0	0	1	1
h_2	0	0	0	1	0	1	1
h_3	0	0	0	0	1	1	1
o_1	0	0	0	0	0	1	0
o_2	0	0	0	0	0	0	1

	i_1	i_2	h_1	h_2	h_3	o_1	o_2
i_1	0	0	1	1	1	0	0
i_2	0	0	1	1	1	0	0
h_1	1	1	0	0	0	1	1
h_2	1	1	0	0	0	1	1
h_3	1	1	0	0	0	1	1
o_1	0	0	1	1	1	0	0
o_2	0	0	1	1	1	0	0

Fig. 5.35 Feedforward ANN with self-aligned recurrent loops of the neurons (self-recurrence, left) and externally aligned recurrent loops (indirect recurrence, right) [47]

To give systems an even higher degree of freedom in training, it is still possible to connect neurons within a hidden layer. This results in a direct competition between the neurons. In contrast to self-recurrence training, neurons inhibit others within the hidden layers to strengthen themselves and achieve an activation. As a result of this direct confrontation, only the strongest neurons succeed in activating, so that they prevail over the others according to the "Winner takes it all" principle. This form of topology is called "Lateral Recurrence" and is shown in Fig. 5.36 (left).

In contrast to the topology options presented, the "Completely Linked Networks" are at the outermost limit. These allow connections between all neurons. This leads to each neuron becoming an input variable, which is the reason why self-linking is no longer possible. As a result, previously defined levels can no longer be clearly separated from each other. In terms of training, this form of topology requires the highest effort, but depending on the application, it can quickly lead to a system being overtrained and, as a result, losing its ability to interpolate and extrapolate into unknown regions of a data set.

Especially at the stock exchange RNNs are widely used, because they are ideal for forecasting time-dependent sequences based on past scenarios. At the forefront of such an application is always the choice of a suitable network architecture and training of this network based on historical data. From now on, the method allows predicting a discrete time P_1 in the future (future prediction) based on the number of past time steps N_p of an

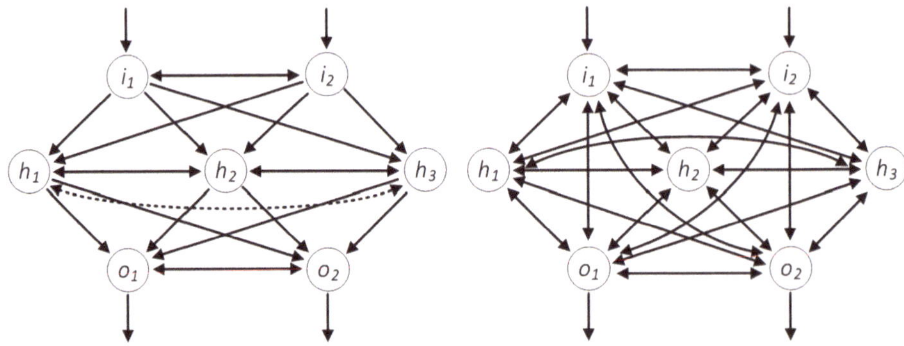

	i_1	i_2	h_1	h_2	h_3	o_1	o_2
i_1	0	1	1	1	1	0	0
i_2	1	0	1	1	1	0	0
h_1	0	0	0	1	1	1	1
h_2	0	0	1	0	1	1	1
h_3	0	0	1	1	0	1	1
o_1	0	0	0	0	0	0	1
o_2	0	0	0	0	0	1	0

	i_1	i_2	h_1	h_2	h_3	o_1	o_2
i_1	0	1	1	1	1	1	1
i_2	1	0	1	1	1	1	1
h_1	1	1	0	1	1	1	1
h_2	1	1	1	0	1	1	1
h_3	1	1	1	1	0	1	1
o_1	1	1	1	1	1	0	1
o_2	1	1	1	1	1	1	0

Fig. 5.36 Feedforward ANN with recurrent loops only within the layers (lateral recurrence, left), completely linked network (right) [47]

experience window (memory window). Starting from the current point in time, this time lies in the future by N_f time steps. The greater the value N_f is chosen, the greater is the probability that the prediction will deviate from reality. The loss can be counteracted by selecting a larger memory window N_p or by increasing the size of the RNN and training it again (Fig. 5.37).

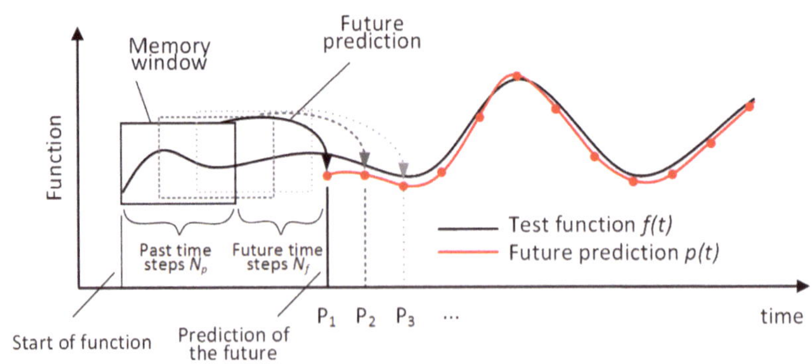

Fig. 5.37 Prediction of time-dependent variables: Definition of past time step and future time step

In addition to recurrent networks, there is a large number of other network structures that have recurring loops and are therefore particularly suitable for data that contains time sequences. These include the Long Short-Term Memory (LSTM) network or the Gate Recurrent Unit (GRU) network. Both types of networks are suitable for classifying, processing and predicting scenarios that lie in the future. Similar to recurrent networks, LSTMs and GRUs are considered deep learning methods due to their complexity and network depth (see Chap. 5.6).

RNNs can be perfectly applied for time-dependent sequences, for which individual RNN networks each consisting of a single neuron are combined to form a chain of modules. This gives them the character of sequential processing. Each module contains two weights, one anchored in the feedforward direction and one in the recurrent loop, which are usually activated by a tanh function (Fig. 5.38).

RNNs are capable of learning dependencies from the past and projecting them to a future point in time. In practice, however, studies by Hochreiter et al. [49] have indicated difficulties that may occur through the simplicity of the network architecture. LSTM networks are a more specialized form of RNNs that are capable of learning and predicting even long-term time dependencies. They were first introduced by the German computer scientists S. Hochreiter and J. Schmidhuber. Their applications are today widespread around the world. LSTMs also have this chain-like structure, but with a more complex processing unit, called LSTM cell. Instead of a single neural network layer, there are four of them interacting with each other through different operators: Two successive layers with a sigmoid activation function, a third layer with a tanh activation and another sigmoid layer (Fig. 5.39).

The key to LSTMs is the cell state $c_{(t-1)}$ (long-term state), which is the horizontal line that runs down the entire chain of LSTM cells in the upper part of the figure, being continuously updated through linear interactions. The sigmoid layers describe how much of each component should be allowed to pass, hence, they are controlling and protecting the cell state. This is why they are also referred to as gate controllers. Their outputs span values from 0 to 1, i.e. they close the corresponding gate with an output of 0, and open it with an output of 1.

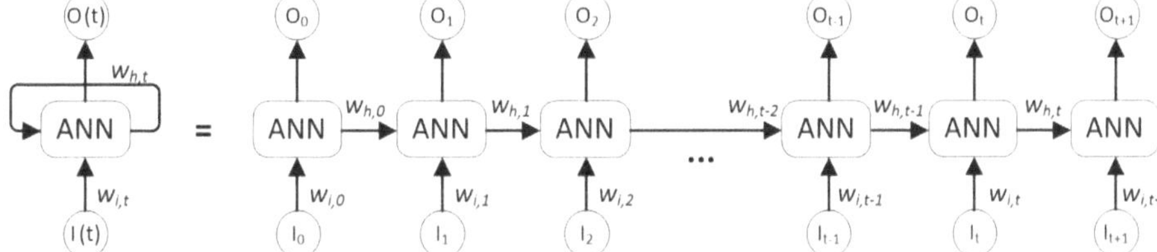

Fig. 5.38 Merging sequential RNNs for processing time-based events

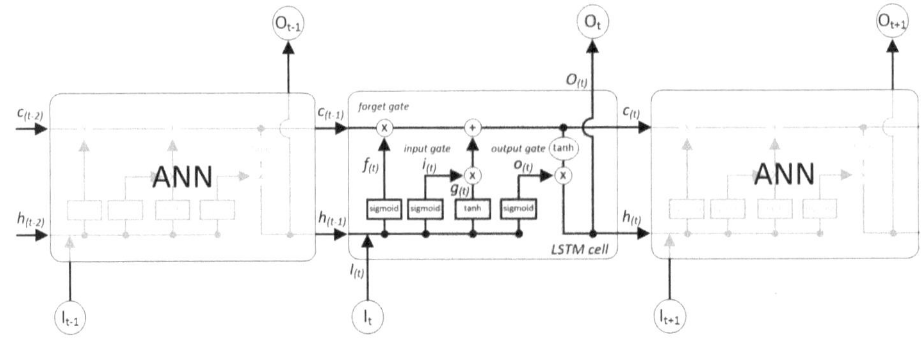

Fig. 5.39 Functionality of an LSTM cell [50]

More precisely, $c_{(t-1)}$ passes through a so-called forget gate $f_{(t)}$. This gate decides which parts of the long-term state are to be deleted. The input gate $i_{(t)}$, on the other hand, controls which parts of the tanh function $g_{(t)}$ should be overwritten and added to the long-term state. Finally, after another operation with the output gate $o_{(t)}$, it is determined which parts of the long-term state should be transported to the outside as a signal. The result is the short-term state $h_{(t)}$ which corresponds to the output of the cell at the time $O_{(t)}$.

In summary, an LSTM cell can learn to recognize an important input, store it in its long-term state, retain it as long as it is necessary and extract it again when needed. This explains why these cells perform amazingly well in capturing long-term patterns of time series.

The following equations summarize how the long-term state $c_{(t)}$ and the short-term state $h_{(t)}$ are calculated. Here, $w_{i,f}$, $w_{i,i}$ and $w_{i,g}$ represent the feedforward weights of the respective four layers to the input vector $i_{(t)}$. The weights $w_{h,f}$, $w_{h,i}$, $w_{h,g}$ and $w_{h,o}$, on the other hand, form recurrent loops back to the short time state $h_{(t-1)}$, (see Fig. 5.38). The terms b_f, b_i, b_g and b_o form the associated bias [50]:

$$
\begin{aligned}
f_{(t)} &= \sigma(w_{i,f} \cdot i_{(t)} + w_{h,f} \cdot h_{(t-1)} + b_f) \\
i_{(t)} &= \sigma(w_{i,i} \cdot i_{(t)} + w_{h,i} \cdot h_{(t-1)} + b_i) \\
g_{(t)} &= tanh(w_{i,g} \cdot i_{(t)} + w_{h,g} \cdot h_{(t-1)} + b_g) \\
o_{(t)} &= \sigma(w_{i,o} \cdot i_{(t)} + w_{h,o} \cdot h_{(t-1)} + b_o) \\
c_{(t)} &= f_{(t)} \cdot c_{(t-1)} + i_{(t)} \cdot g_{(t)} \\
O_{(t)} &= h_{(t)} = o_{(t)} \cdot tanh(c_{(t)})
\end{aligned}
\tag{5.9}
$$

Example 10: Predicting Emissions from the Real Driving Emission Cycle (RDE)

Emissions in real driving conditions are described as real driving emissions (RDE), unlike those measured under non-real test conditions on an exhaust roller dynamometer. Due to the fact that in real driving conditions, passenger cars certainly emit more exhaust gases than in the new European driving cycle (NEDC) test procedure, which has been used for vehicle registration in Europe since 1992, the European Union decided in 2015 to supplement the test procedure by the RDE test. The test procedure has been effective since September 2017 for cars, trucks and buses in everyday use. The great challenge of this test method is that, unlike other driving cycles, the distance and speed as well as the elevation profile are not fixed. The route that is driven during the RDE test consists of an urban, a rural road and a freeway section. The proportions are kept flexible within predefined limits, and the driving behavior during the test operation is not subject to any fixed specifications and can be individually adjusted by an inspector. These freedoms mean that it is difficult to accurately predict emissions using common models.

In the following example, a self-recurrent RNN is selected (see Fig. 5.35), which consists of 4 hidden layers and 30 neurons each. An RDE test cycle and associated measurements of carbon monoxide (CO) and nitrogen oxides (NO_X) recorded with a portable emissions measurement system (PEMS) serve as the basis for the observation. The first 4300 s of the test drive, representing 75% of the total test time, are used to train the neural network. For the remaining distance, a prediction of emissions shall be made. Alternatively, it is also possible to use a complete RDE measurement as a training basis and, based on this, to predict emissions for other RDE measurements (Fig. 5.40).

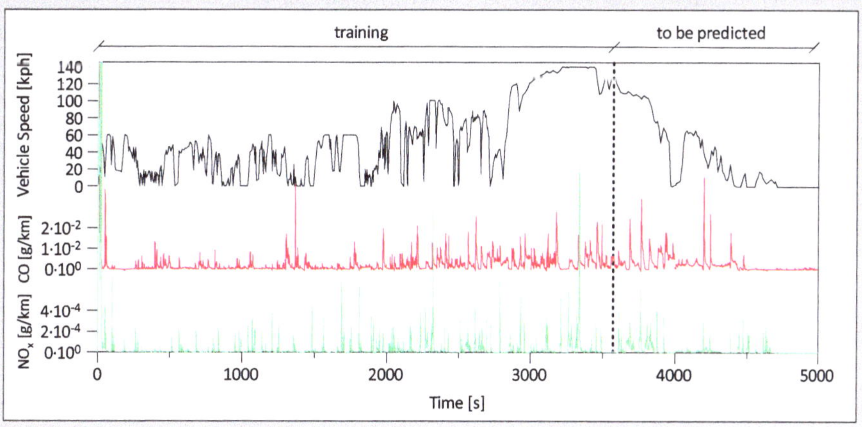

Fig. 5.40 Training and prediction of CO, NO_X emissions of the RDE test procedure

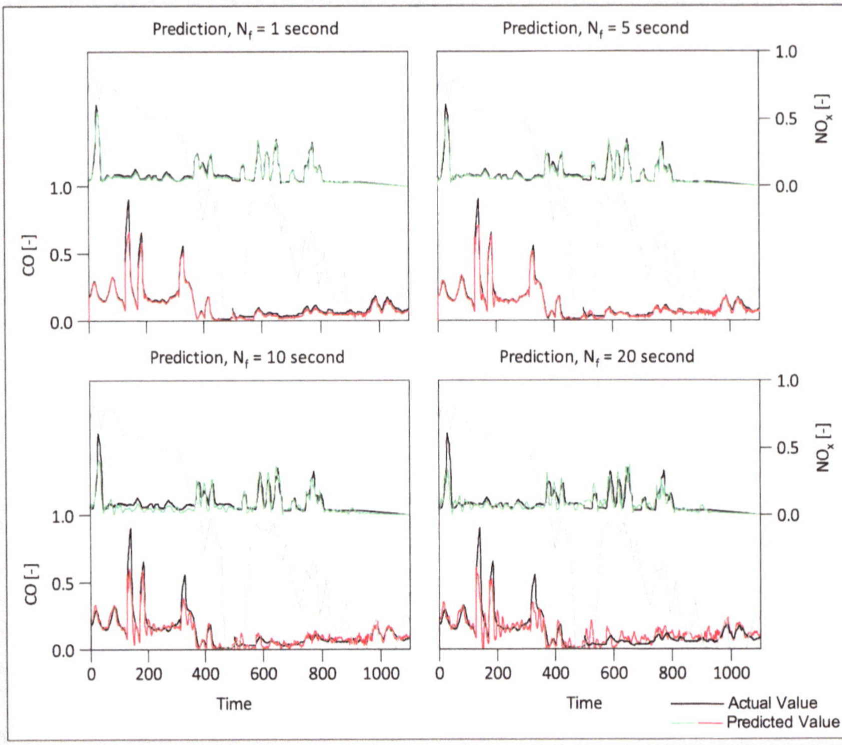

Fig. 5.41 Prediction of CO and NO_x in RDE for different future times steps N_f

When using RNNs, the user has the choice to define the time step size of the prediction. If a prediction lies far in the future (large N_f), it is recommended to counteract with an equally large N_p and/or a larger network architecture; see Sect. 5.5.4. The following graph shows the result of the predicted CO and NO_x on the RDE test for N_p=300 and N_f = 1 second, N_f = 5 seconds, N_f = 10 seconds and N_f = 20 seconds in the future. The quantities were normalized on the basis of the respective maximum value [0-1] (Fig. 5.41).

5.5.3 Training Procedure

Similar to biological neurons, which are inhibited or reinforced by their upstream synaptic connections, weights take over this task in artificial neural networks. During the training phase of an ANN, the weights are always adjusted. For this purpose, different training algorithms have been developed, of which only a few are suitable for multi-layered neural networks. Some of them are suitable for feedforward networks where the information flow is unidirectional, others for recurrent networks where the information flow is backward, as presented in Sect. 5.5.2.

Detached from the nature of the network architecture, there are training procedures which are suitable if the output of the ANN is known (Supervised Learning; see Sect. 5.4.1), procedures for ANN with unknown output variables (Unsupervised Learning; see Sect. 5.4.2) and special procedures for self-learning systems (Reinforcement Learning; see Sect. 5.4.3). Which training method ultimately is the most suitable depends on the respective application.

In this section, some training procedures are presented. For an overview, please refer to the flowchart in Fig. 5.5.

Backward Propagation—Supervised Learning

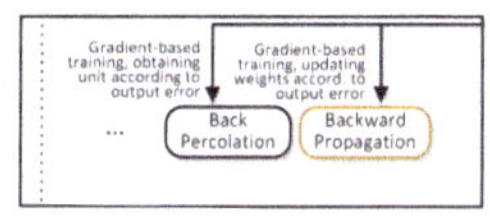

Backward propagation is one of several training methods to train artificial neural networks. The training is a supervised learning method, i.e. data that are subjected to training must be classifiable; see more details in Sect. 5.4.1.

At the beginning of a training process, all weights of a neural network are initialized. A good initialization is of central importance because it determines how many epochs (iterations) a network has to be trained to achieve stability. In addition, the final result can differ if the initialization is bad. One of the most important initialization methods are the ones of G. Xavier and K. He [51, 52].

After this step, the net now generates corresponding output data of the zeroth iteration. A deviation (error) from the expected output is calculated. The error is propagated backwards to the previous level. By adjusting the weights on this layer, the error is reduced by a gradient minimization method (gradient descent method). This process is repeated until the error falls below a certain limit. Once the smallest gradient is found, which corresponds to a global minimum, the training is considered to be completed.

The training process can be divided into three steps and is explained in more detail below using the example of a simple neural network (multi-layer perceptron) with two input signals (x_1, x_2), two hidden layers and an output signal y:

1. Forward Propagation
   Network training is an iterative process. In each iteration, the weights at each node are modified using new training data. Each training step starts with a new feed of input signals from the training set. Afterwards, the output signal values can be determined in each network layer for each neuron. The following pictures illustrate how a signal propagates through the network (Fig. 5.42).

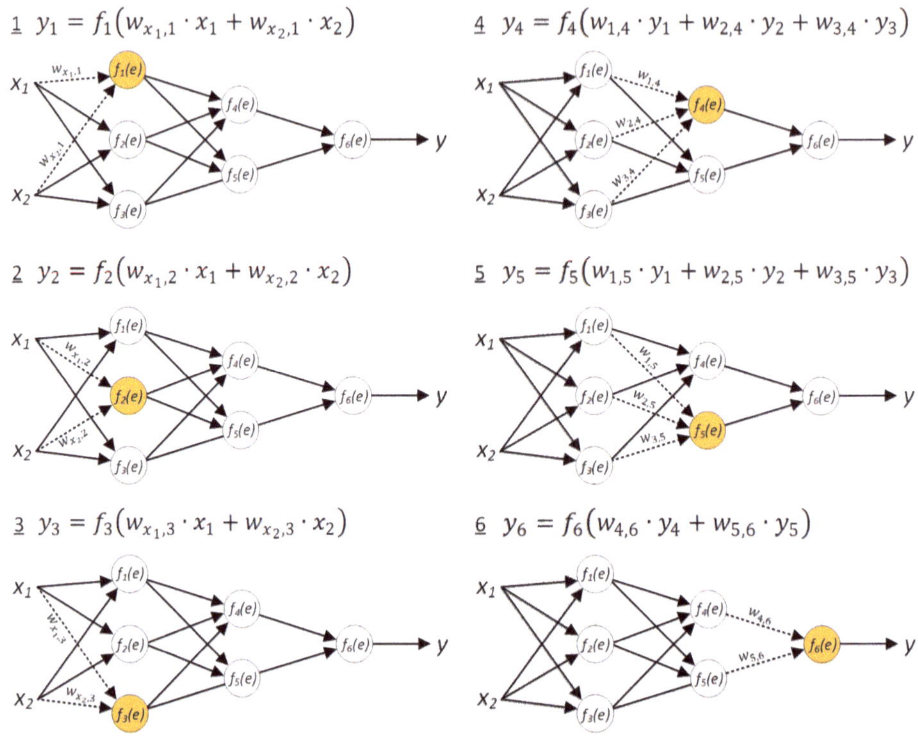

Fig. 5.42 Backward propagation training phase 1: Feedforward and derivation of individual neural outputs [53]

2. Calculation of the errors
   In the second step of the training phase, the output signal y of the network is compared with the target value, which is part of the training data set. The difference $\delta = z - y$ quantifies the error of the network.

   Any cost function C can be selected as a training basis for updating the weights.

$$C = \frac{1}{2} \cdot (z - y)^2 \tag{5.10}$$

   The coefficients $w_{i,j}$ of the weights used to retransmit errors are the same as those used during the calculation of the output value. It is therefore essential to select the same activation function for both the training process and the subsequent calculation process, otherwise the quality of the training results could not be reproduced (Fig. 5.43).

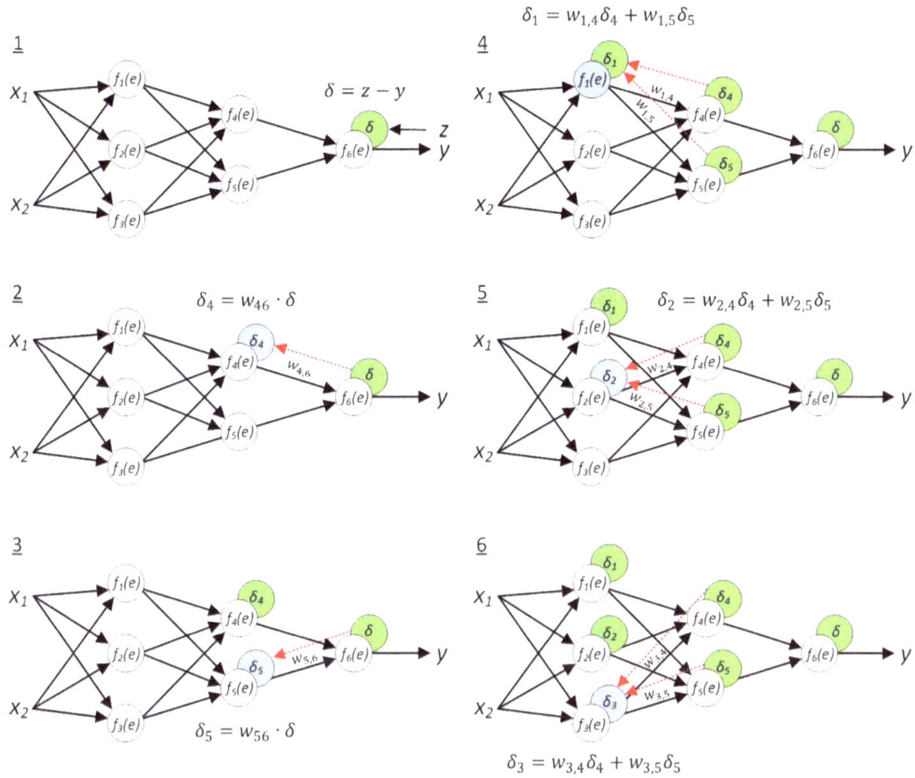

Fig. 5.43 Backward propagation training phase 2: Derivation of errors and gradients [53]

3. Update of the weights
   To adjust the weights, the partial derivative of the error function is performed:

$$\delta_k = \frac{dC}{dw_{i,j}} = (z - y_k)\frac{dy_k}{dw_{i,j}} \tag{5.11}$$

y_k describes the activation function as shown in Fig. 5.32. For the choice of a sigmoid activation function, y_k is

$$y_k = F(s_k) = \frac{1}{1 + e^{s_k}} \tag{5.12}$$

When inserting Eq. 5.12 in Eq. 5.11, the result is

$$\delta_k = \frac{dC}{dw_{i,j}} = (z - y_k) \cdot y_k(1 - y_k) \tag{5.13}$$

After the gradients are calculated, the weights are adjusted:

$$w_{i,j}^{*} = w_{i,j} + \Delta w_{i,j} \tag{5.14}$$

with

$$\Delta w_{i,j} = \eta \cdot \delta_k \cdot y_k \tag{5.15}$$

where η describes the learning rate of the backpropagation process by influencing the training speed in the network. There are two fundamentally different approaches to select these parameters. The first one is to give the training process a large starting value. While the weights are set, the parameter is gradually reduced. The second approach starts by teaching with a small parameter value. During the training, the parameter is increased while progressing and decreased again in the final stage. This makes it possible to determine the signs of the weights faster. This approach takes more time but reacts more robustly to a convergence criterion (Fig. 5.44).

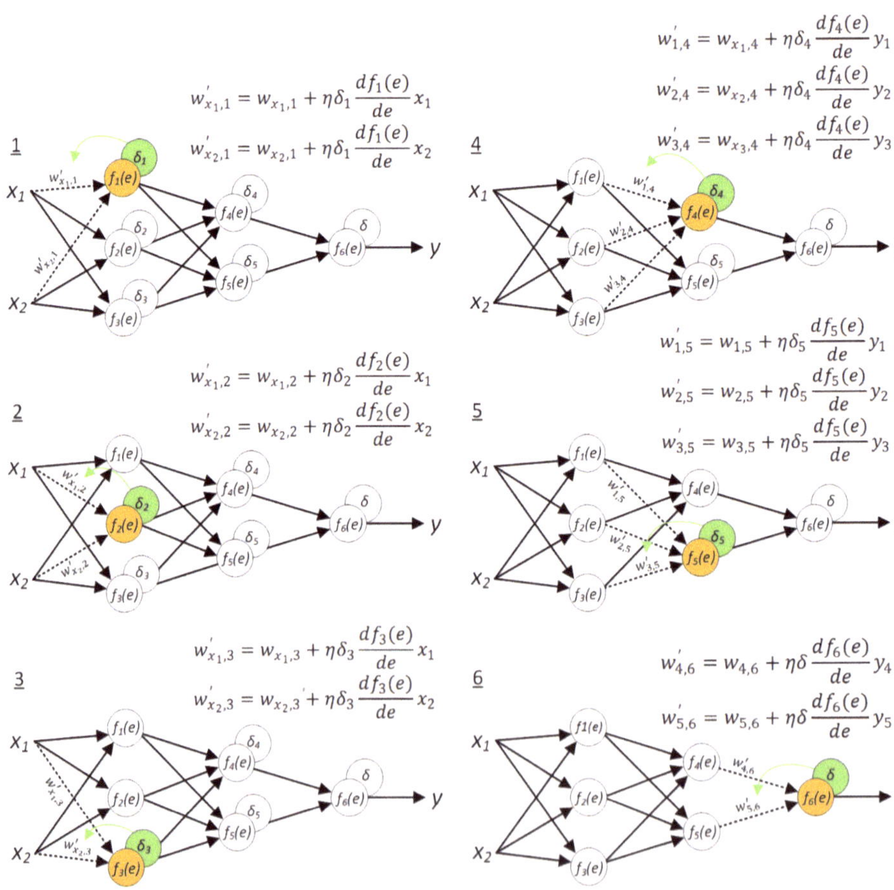

Fig. 5.44 Backward propagation training phase 3: Update of the weights based on the gradients [53]

Counter-Propagation—Unsupervised Learning

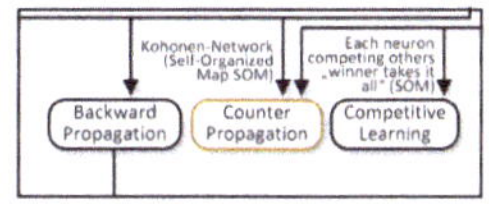

In 1987, the US-American computer scientist Robert Hecht-Nielsen presented a new form of training method for neural networks called "Counter-Propagation" (CPN). In contrast to backpropagation, this is an unsupervised learning method (see Sect. 5.4.2 for more details) and is especially suitable for feedforward networks. The counter-propagation has self-learning properties and is particularly designed for applications where a database is not classifiable, especially when input variables are dynamic, i.e. always changing, and output variables are unknown.

The counter-propagation training is typically combined with a Kohonen hidden layer. Kohonen networks are also called self-organizing maps (SOM), which are mainly used to visualize and analyze highly complex data. Each data point is assigned to a network node according to the similarity theorem and metric evaluation basis. This enables the method to "map" input variables directly to an output. The output layer is called the Grossberg layer.

In principle, a data mapping, i.e. a direct projection of an input signal onto a possible output can also be solved with the backpropagation method. However, the counter-propagation method promises a higher prediction and up to 100 times faster computation speed, especially for dynamic and variable data patterns. Particularly for applications where real-time is required, CPN is well suited.

Figure 5.45 shows a counter-propagation network with five layers, the input layers 1 and 2, a hidden layer 3 and the output layers 4 and 5. With a given vector pairing $\{x_1, y_1\}$, $\{x_2, y_2\}, \ldots, \{x_n, y_m\}$ the CPN tries to correlate an input vector $\{x_i\}$ with an input vector $\{y_i\}$ during the training phase. If a correlation can be described with a continuous function Φ, so that $y = \Phi(x)$ is true, the CPN learns to classify this solution for each vector $\{x\}$ of the training space of the same characteristic. If an inverse correlation is detected such that the vector $\{x\}$ is also a function of $\{y\}$ by $x = (\Phi - 1) \cdot y$, the CPN learns to map inversely. In this case, it is a "full counter-propagation network", otherwise it is a " forward-only counter-propagation network".

The training concept is based on a two-step process. First, an unsupervised learning method is applied. The Kohonen layer (Hidden Layer) works according to the "Winner takes it all" principle. According to this principle, the strongest neuron prevails in competition with the others by exceeding the threshold value (bias) and thus gets activated. The Kohonen level literally provides a classification. The activated node carries out a "supervised learning" method in the Grossberg layer (output level) and generates the output signal. In this respect, the training method is also categorized as semi-supervised learning.

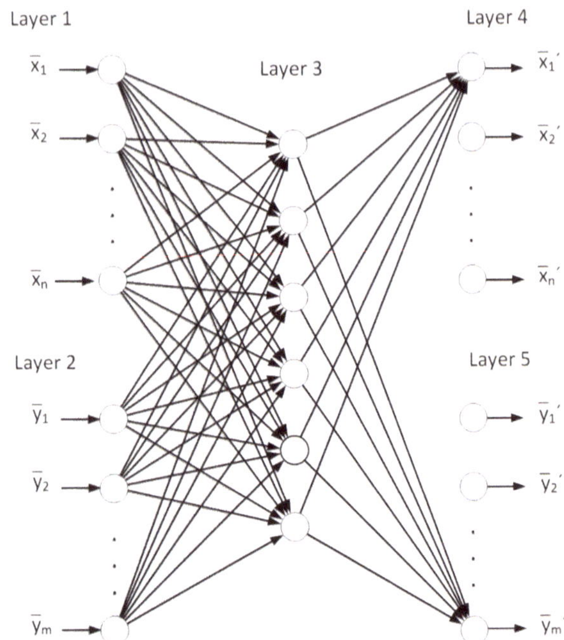

Fig. 5.45 Neural network for counter-propagation training [54]

Competitive Learning—Unsupervised Learning

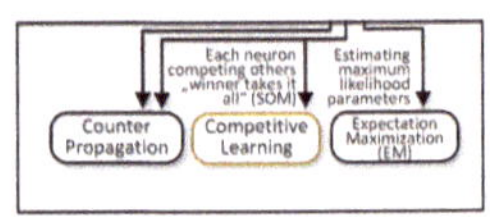

Among the best known competitive learning concepts are self-organized networks (Self-Organizing Maps: SOM). SOMs represent another form of artificial neural networks, which were presented in 1982 by the Finnish engineer Teuvo Kohonen. They are one of the most prominent unsupervised learning methods, which aim to reduce and discretize the dimension of complex data inputs to a 2D output plane (2D map). SOMs differ from common ANNs as the training of the weights is not performed by a gradient descent method (backward propagation), but by competitive learning. Each neuron works "self-organized", i.e. it competes independently with its neighboring neurons, which is according to the "Winner takes it all" principle. The weight of a neuron is rewarded if it produces results that are most similar to a randomly selected vector.

Based on common signal processing concepts, a so-called quantization error (QE) provides a measure of the average distance between the data points and the map nodes on which they are mapped. Kohonen proposed QE as a basic quality measure for the evaluation of self-organizing maps. The value for a map is calculated using the following equation, where n represents the number of data points in the training data and $\Phi : D \rightarrow M$ represents the projection of the input space D onto the feature map M of the SOM [55]:

$$QE(M) = \frac{1}{n}\sum_{i=1}^{n} \|\Phi(x_i) - x_i\| \tag{5.16}$$

Another goal of the SOM algorithm is to preserve the topological features of the input space in the output space. For this purpose, there is another important measure called topographic error (TE) that describes how well the structure of an input space is represented by the map. TE calculates for each input the best and second best matching neurons of the map and based on this, evaluates its position. If the nodes are adjacent, it means that the topology for that input has been obtained. Otherwise, this is evaluated as an error. The total number of errors divided by the total number of data points results in the topographic error of the map, where $\mu(x)$ denotes the best matching value for the data point x and $\mu^{'}(x)$ denotes the second best [55]:

$$TE(M) = \frac{1}{n}\sum_{i=1}^{n} t(x_i) \tag{5.17}$$

$$t(x) = \begin{cases} 0 & \text{if } \mu(x) \text{ and } \mu^{'}(x) \text{ are neighbors} \\ 1 & \text{otherwise} \end{cases}$$

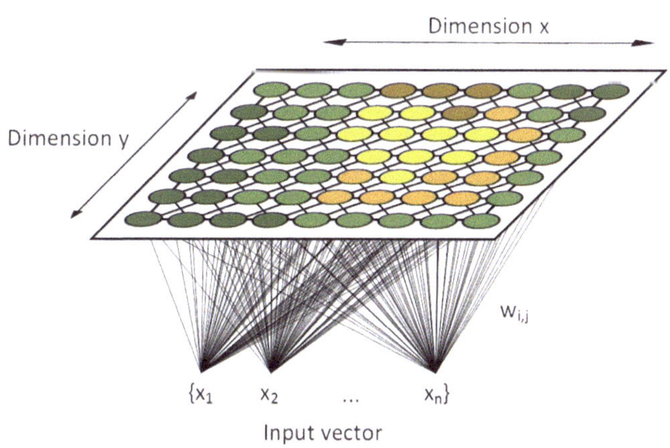

Fig. 5.46 Projection of data onto a 2D feature map (Self-Organized Maps: SOM) [56]

Each neuron acquires its own x-y coordinate on the feature map (color map). Through the simultaneous training of the neighboring neurons, the feature map grows gradually, which is the result of the clustering process. In this respect, each data point is assigned to a certain characteristic by representing a color class (Fig. 5.46).

MDP, Q-Learning, SARSA

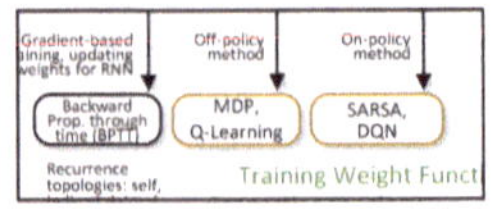

1. To describe the reaction behavior to environmental influences, so-called Markov decision processes (MDPs) can be used as a general framework for reinforcement learning. An MDP consists of a set of finite environmental states s, a set of possible actions in each state $a(s)$ and a reward function $r(s)$. The goal of an MDP is to calculate the optimal policy $\pi(s, a)$ so that a corresponding action can be taken for any possible state. This action is being rewarded, if it brings a success. The goal is to maximize the rewards as much as possible. However, since real environments can be highly dynamic, it is difficult to predict their reactions to changing input data. In such cases, the use of model-free RL methods is more appropriate, since they are better suited to predict unexpected effects.

2. Q-learning and State-Action-Reward-State-Action (SARSA) are two frequently used model-free RL algorithms. While Q-learning is an off-policy method in which the agent learns its environment based on a policy, SARSA is an on-policy method that learns it mainly based on its current action and secondly by the help of additional policies.

The weight for a step from a state t by Δt steps into the future is calculated with $\gamma^{\Delta t}$, where for γ $(0 \leq \gamma \geq 1)$ applies. This has the effect that earlier received rewards are valued higher than rewards received later. γ can also be interpreted as the probability of success after each time step Δt. Before the training begins, Q is initialized to any random value. Then the agent selects an action a_t at any time t, observes the reward r_t and assigns a new state $s_t + 1$. After this, Q is updated [57]. Both Q-learning and SARSA are based on the following equation:

$$\underbrace{Q(s_t, a_t)}_{\text{new value}} \overset{\text{updating direction}}{\longleftarrow} \underbrace{Q(s_t, a_t)}_{\text{old value}} + \overbrace{\alpha}^{\text{learning rate}} \cdot \overbrace{(\underbrace{\underbrace{r_t}_{\text{reward}} + \underbrace{\gamma}_{\text{discount factor}} \cdot \underbrace{max\, Q(s_{t+1}, a)}_{\text{estimate optimal future value}}}_{\text{new value}} - \underbrace{Q(s_t, a_t)}_{\text{old value}})}^{\text{temporal difference between current and learned value}} \tag{5.18}$$

The methods are easy to implement but not generally applicable, because they are not able to estimate values for unknown states. This can be overcome by more advanced algorithms like deep Q-networks (DQN), which use neural networks to estimate Q-values. However, DQNs are more suitable for processing discrete and low-dimensional action spaces.

Reinforcement learning is widely used where training is based on variable input data and unpredictable output data, such as in the area of autonomous driving. The environmental influences that can occur while driving are complex and cannot be defined finitely. The situation is similarly complex in the field of "gaming", where virtual game partners are developed, or in the field of robotics.

5.5.4 Network Architecture and Performance

The architecture of a neural network describes its composition or dimension and is defined by the number of hidden layers (HL) and the number of neurons per hidden layer (Hidden Units: HU) that the network comprises. At the beginning of modeling, a network, searching for the optimal architecture, is one of the key aspects the user usually has to deal with.

The determination can be very time-consuming, therefore it is recommended to follow some general guidelines in order to get a possible orientation in advance. In principle, the optimal architecture is strongly related to the number of input variables of the model and the size of the data set used for training the network. Fig. 5.47 illustrates the dependency of the network architecture on its performance according to the amount of training data. Performance can also be understood as the capability of prediction, which can be expressed analytically either by the coefficient of determination (R^2) or by other standard evaluation quantities like mean square error (MSE) or root mean square error (RMSE).

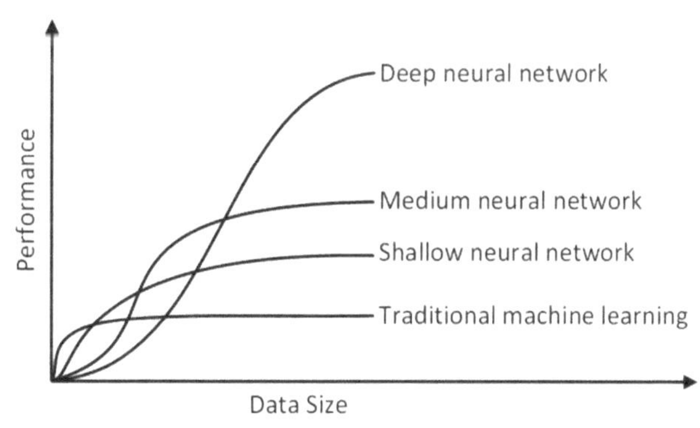

Fig. 5.47 Performance of an ANN as a function of data volume and network architecture [58]

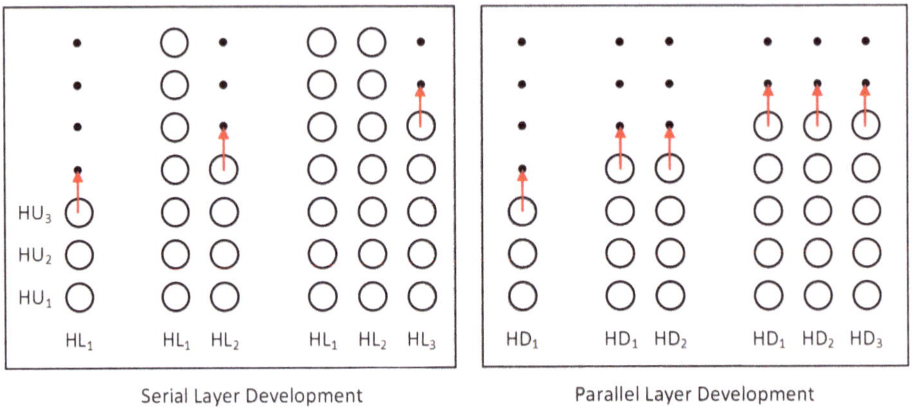

Fig. 5.48 Sequential construction of a neural network according to a serial method (left) and parallel method (right)

There are strategically different approaches to building a network. In the following, two sequential methods are presented: 1. a serial developing structure of HLs and 2. a parallel developing structure of HLs. For both methods, a maximum number of HUs must first be defined. In the serial setup, the HUs are developed one after the other, see Fig. 5.48 (left), while the performance of the network is always observed. When an HL is fully developed, the construction of the next HL level begins. In parallel setup (right), however, the number of HUs per HL is set up anew for each newly added HL. In principle, it is also possible to variably generate the number of HUs for each HL. This results in a permutative number of possible combinations, which leads to an exponential increase of arithmetic operations, which is why it is inadvisable.

Using the example of a set of 45000 training data points, 8 inputs and one output variable, the performance of a network during its development is shown. For this purpose, 75% of the data is used for training and 25% for validation. The coefficient of determination R^2 is shown as the evaluation variable, which represents the correlation between the real data and the results calculated from the model. For this example, a serial network reaches its optimal performance at about 3 fully trained HLs and 100 HUs. In contrast, the parallel development method finds an optimum at 4 HLs with 80 HUs. In principle, one can say that both methods propose similar architectures. Depending on the problem and the existing knowledge of a system, one or the other method can result in time advantages. In the search for an efficient network architecture, both methods should be weighed against each other (Fig. 5.49).

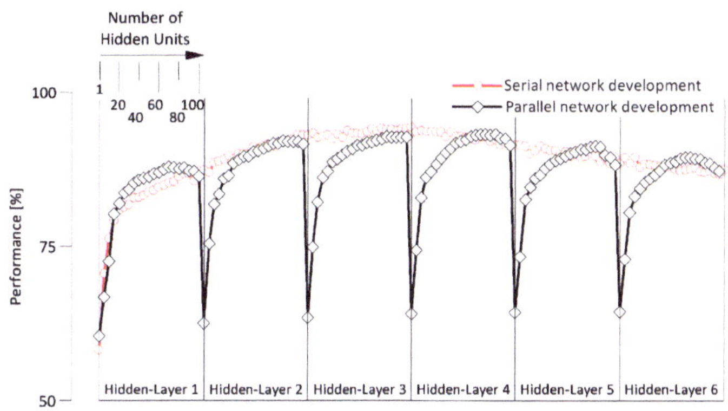

Fig. 5.49 Training of a serial and parallel network

The number of HUs reacts much more sensitively than the number of HLs for the increase in performance. The focus should therefore be more on the development of neurons than on a large number of HLs. As the number of HLs increases, the additional gain in network performance becomes smaller and smaller (Fig. 5.50).

Fig. 5.50 shows that the first HL is (are) considered the performance-defining pillar(s). If further HL levels are added, smaller potentials can be localized for large data sets (improving pillars). Any further levels from there represent only a slight increase in performance and

Fig. 5.50 Network dimension versus efficiency

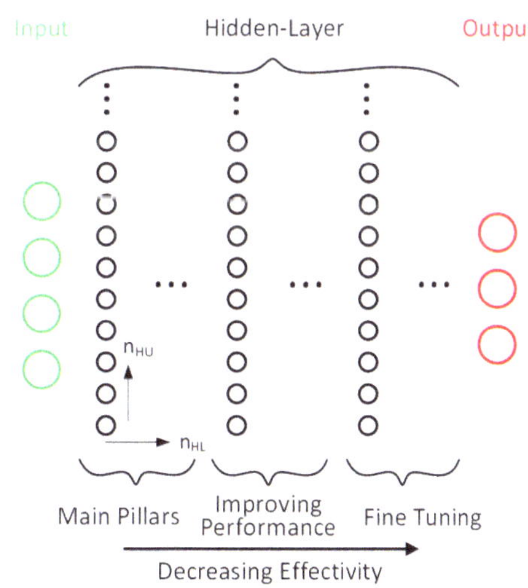

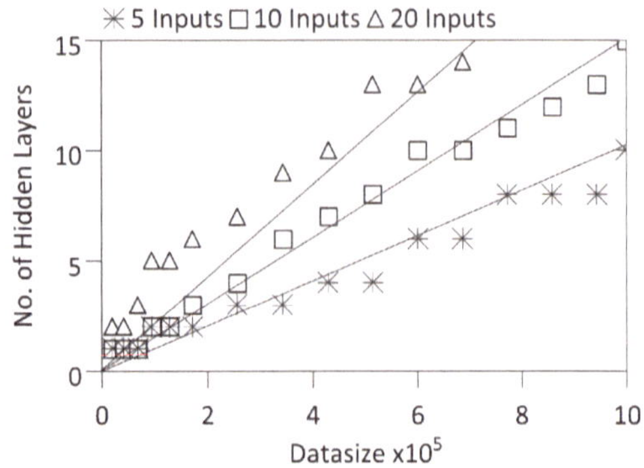

Fig. 5.51 Number of hidden layers to be selected depending on data volume and number of input variables

thus are identified as "fine-tuning pillars". The choice of a network structure is one of the sensitive issues, as it has an immense influence on the efficiency of a model. At the same time, sequentially enlarging the architecture according to Fig. 5.48 and approaching an optimum can be very time-consuming. In order to help the user to select a suitable architecture, the following illustration provides a rough orientation for the selection of the HL number. Depending on the data size and the number of input parameters, it can be described linearly (Fig. 5.51).

If the HL number was defined in the first step, the number of HUs remains as a further unknown variable. In the second step, it can be roughly selected for 5, 10 and 20 input variables using Fig. 5.52.

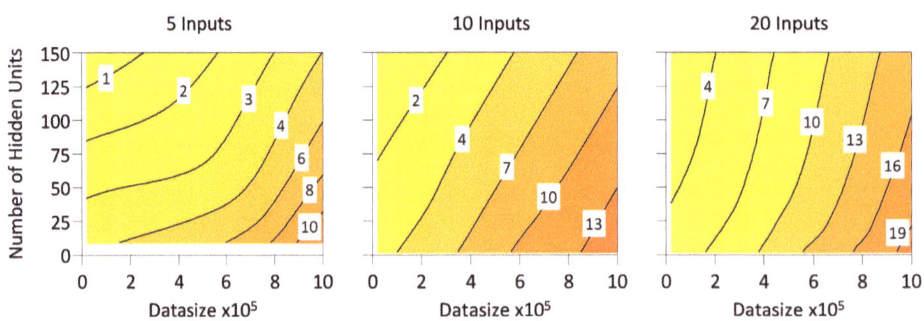

Fig. 5.52 Number of hidden units to be selected depending on the amount of data

Hyperparameter

In machine learning applications, there are a number of sensitive parameters that have a relevant effect on the performance of a neural network. Those that are used to control training and whose value, unlike other parameters, must be determined before a model is trained are called hyperparameters. The dimension of an ANN is therefore also counted as a hyperparameter.

The most important hyperparameters, which in sum have a decisive influence on the performance of an ANN, are listed as follows:

1. **Number of hidden layers and hidden units**
   As explained in the previous section, the dimension of an ANN plays a crucial role in the creation of a suitable model. The larger and more complex a data set and the more input variables it contains, the more likely a performance optimum of the ANN will shift toward a higher dimension.
2. **Dropout**
   Dropout is a technique used to increase network performance and avoid overfitting. By using a trial and error method, individual neurons are canceled out of the architecture, while the performance of the network is constantly checked. A dropout value of 20%–50% of neurons is recommended as a guideline. Especially for large networks, this method promises to uncover considerable potentials, and speed up the simulation time (Fig. 5.53).
3. **Activation Function**
   Activation functions are used to transfer nonlinear relationships of data into models. Depending on which characteristics and properties a data set has, it can make qualitative differences with which activation function a corresponding model is trained. As far as it has to be selected in advance of training, it is one of the most important hyperparameters. Further details were discussed in Sect. 5.5.1.

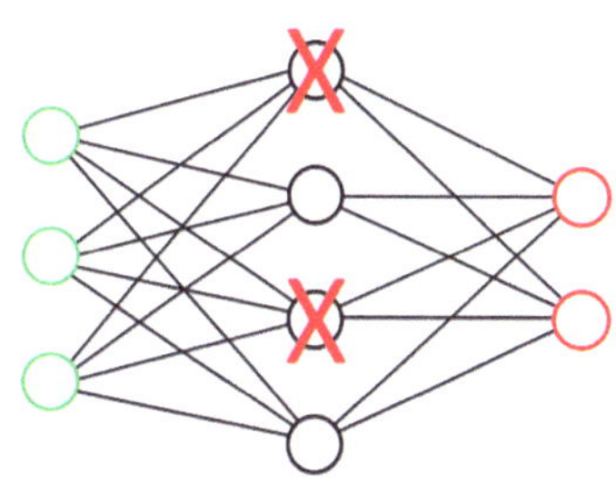

Fig. 5.53 Dropout of neurons [59]

4. **Weight Initialization**
   A sensible initialization of the weights at the beginning of a training decides on how many epochs a network has to be trained to reach its maximum performance. In addition, a good initialization can achieve a better training result than an unfavorable initialization. One of the most important initialization methods is that of G. Xavier and K. He, who estimate the best possible initial values of the weights based on an upstream analysis of the respective data [51, 52].
5. **Train versus Test Ratio**
   For a given data set before a training session starts, it is common practice to specify what proportion of the data set will be used for the training itself and which remaining part is used for the validation of the model (test). According to different sources, a ratio of 70%–85% is recommended between a training and a test data set. The selection of the portions shall be random. This value can be sensitive to the performance.
6. **Batch Size**
   If a data set is considerably large, the training process can result in an enormously high computing load. Especially for studies where a process is recurrent, progress can be inefficient in terms of time. One method that helps to reduce training time is to segment a data set into small packages (batches). Instead of feeding the entire data set once, the batches are then transferred sequentially to the network. The weights are updated after each input of a new batch. A default value for the batch size is 32, and for larger batches multiples of this number can be selected (Fig. 5.54).
7. **Learning Rate**
   Learning rate determines how fast the weights of a network should be updated. A low learning rate slows down the learning process, but is robust with respect to oscillations and more stable with respect to convergence. A higher learning rate can speed up the learning process significantly but runs the risk of oscillating. This parameter is typically chosen depending on the batch size. The higher the data set or the batch size, the more reasonable it is to choose a higher learning rate to counteract the slow learning process. It is also common to set a learning rate that degrades during the training. If a plateau is reached during optimization, the reduction of the learning rate can lead to higher robustness and initiate a fine-tuning process.

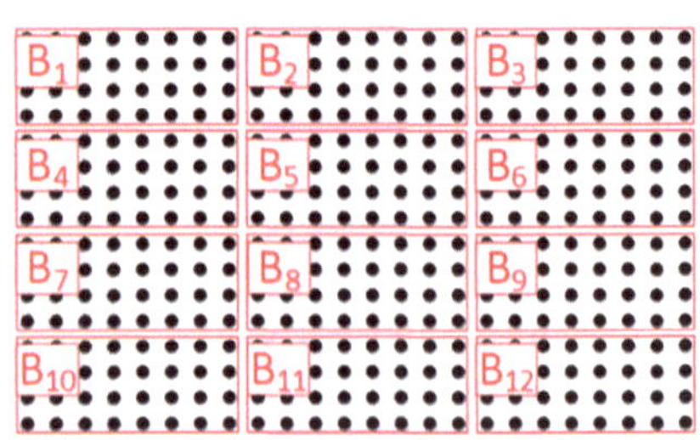

Fig. 5.54 Subdivision of a data set into batches

Fig. 5.55 Training/Test versus epochs

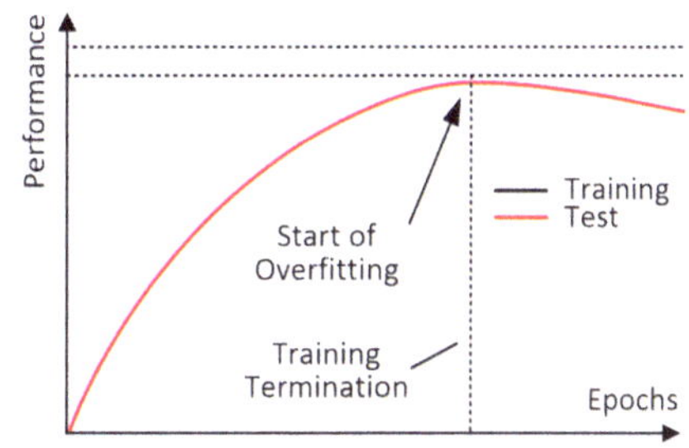

8. **Number of Epochs**
   The iteration of a training step (see Figs. 5.42, 5.43 and 5.44 using the example of the backward propagation procedure) is called an epoch. As a rule, so many epochs should be performed until the validation quality of the test data reaches a maximum. The training process should therefore be terminated by a termination criterion. If the training accuracy increases during further training while the validation accuracy of the test data remains the same or even degrades, it is a sign that overfitting is taking place (Fig. 5.55).

A manual determination of optimal hyperparameters by a trial and error method is enormously time-consuming and inefficient. Due to constantly growing challenges, AI users now have modern and reliable algorithms called Automated Machine Learning (AutoML) at their disposal, which highly automates upstream screens and the search for hyperparameters. The degree of automation promises to adapt the techniques of machine learning on a high level even to non-experts.

AutoML algorithms support from the input of raw data all the way to the complete development of predictive models and help users to shift their tasks more and more to overarching topics rather than having to deal with detailed tasks such as pre-processing of data, feature extraction and feature selection, algorithm selection and hyperparameter optimization. AutoML algorithms use optimization methods such as the Manual Search, Grid Search, Random Search and the Bayesian Optimization, which will not be discussed in detail here.

5.5.5 Fuzzy Logik

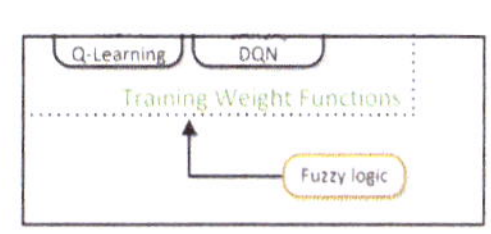

The term fuzzy logic was first published in 1965 by the Azerbaijani mathematician and computer scientist Lotfi Aliasker Zadeh at the University of California, Berkley. The basis for this, however, lay in much earlier studies of the1920s by Jan Łukasiewicz and Alfred Tarski known through the so-called Łukasiewicz logic.

This topic deals with the handling of data that are subject to a certain fuzziness. In contrast to the concept of Boolean logic, where quantities are rounded to 0 and 1 to categorize fuzzy quantities, fuzzy logic allows assigning any value between 0 and 1 to a quantity, which is the basis of the concept of partial truth.

Fuzzy logic is based on observations of human decision behavior. Often these are made on an inaccurate and non-numerical background information. On the one hand, the information can be strongly founded and validated or it can be based on a conglomerate of unconscious experience values, which manifest themselves as a so-called "gut feeling". The intermediate stages are vague and not differentiable—this is exactly the area that fuzzy logic tries to represent numerically and quantify on a mathematical basis. The concept of fuzzy logic has a wide range of applications, especially in the field of control (electronics) and artificial intelligence.

Using a simple example of the subjective sensation of test persons who feel a liquid of different temperatures (in this case water) on their skin, the following diagram gives an example of how statistical sensation can be classified into individual categories using fuzzy logic. It can be seen from this that when control strategies of this type of subjective parameters are used, a range is obtained whose weighting is taken into account by fuzzy logic. In this example, 72% of the test persons perceive a water temperature of 39° C as hot and 28% as warm. In reality, dependencies cannot be represented by simple, linear relationships, as shown in Fig. 5.56.

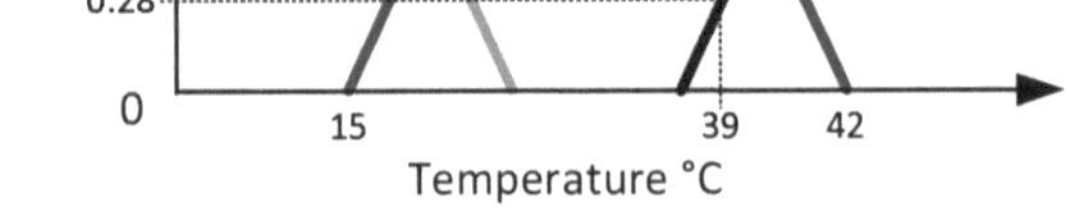

Fig. 5.56 Temperature sensation in a group of test persons

Combination of fuzzy logic and artificial neural networks

In the past, intelligent, hybrid systems combining fuzzy logic and neural networks have proven their high efficiency in a variety of real-world application problems. While neural networks, for example, are predestined to recognize patterns from a large amount of data and henceforth to expand their model predictivity in a self-learning manner, their decision-making is fundamentally inexplicable or incomprehensible. Fuzzy logic, on the other hand, can argue with the help of inaccurate information and present decisions in a comprehensible way, but it cannot determine the rules it uses for these decisions. As a result of the two-sided limitation, the motivation to combine both systems into an intelligent hybrid system becomes apparent. The computational process intended for fuzzy neural systems can be divided into the following three steps:

- The first step is the development of a "fuzzy neuron" based on the understanding of biological neural morphologies and subsequent learning mechanisms.
- This is followed by the integration of fuzzy neural models with synaptic connections, which include a fuzziness in the neural network.
- Finally, the development of learning algorithms for the adaptation of synaptic weights follows.

Basically, there are two possible ways to design fuzzy neural systems:

1. A multi-layer neural network controls the fuzzy mechanism; see Fig. 5.57 (left).
2. The fuzzy interface provides an input vector for a multi-layer neural network. The neural network is adaptive and is trained to calculate desired command outputs or decisions; see Fig. 5.57 (right).

Fig. 5.57 Combination of fuzzy logic and a neural network as a hybrid system for 1. ANN as data supplier (left) and 2. fuzzy logic as data supplier (right) [60]

Example 11: Subjective driving behavior for the calibration of electronic control units

By definition, subjectivity describes the relationship of a subject to its environment. In a derived sense, a customer's decision to purchase a vehicle is based on subjective driving impressions. Particularly noteworthy here is the so-called "drivability", which is the sum of many individual technical components at the end of vehicle development. However, even the application engineer who is responsible for the application of electronic control units can influence the drivability of a vehicle to a certain extent.

For many years, many central research projects have been dealing with the development of suitable calculation methods for test procedures in order to transform subjective driving experience patterns into objective evaluation criteria. Finding a clear match is very complex, especially because a driving experience can vary according to country of origin, road conditions, age, gender and other factors. In [61], *an approach is presented where objective parameters for the drivability are divided into 3 categories: 1. suspension comfort, 2. steering response and 3. load change behavior. They have proven to show a good agreement with subjective parameters. For a multitude of vehicle classes and subjective measurement data of drivers, the parameters provide information about trigger points in the vehicle, so that optimizing measures can be initiated concretely.*

If the three categories (suspension comfort, steering Response and load change behavior) are divided into 5 evaluation levels, a very useful application for the hybrid combination of fuzzy logic results: (Fig. 5.58*)*

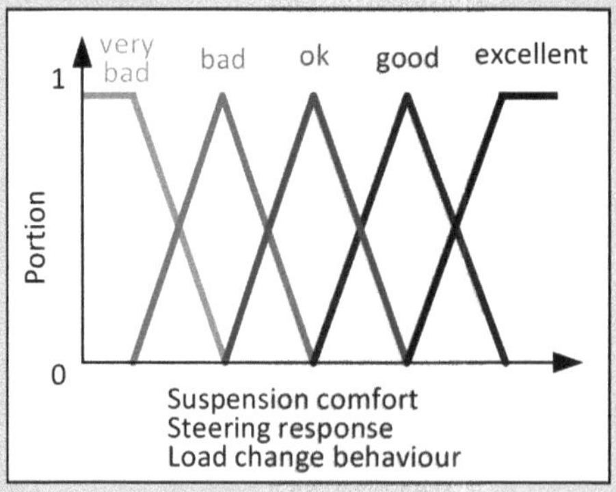

Fig. 5.58 Weighting of different load shares of a driving cycle using fuzzy logic

5.6 Deep Learning

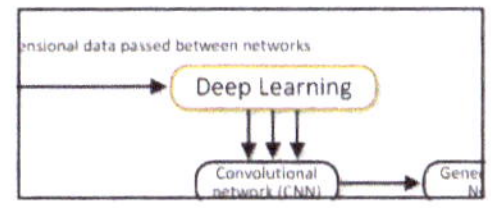

Deep learning is a special learning method for artificial neural networks, which is characterized by a much more complex structure than the networks discussed so far. Basically, an ANN is also called Deep Neural Network (DNN) by definition, if it has at least two hidden-layer levels. Besides DNNs, the term deep learning also includes other forms of networks such as the convolutional neural network (CNN), which is mainly used for image and sound recognition [62].

For neuronal networks of the category deep learning, multiple learning methods can be used: Classification and regression of known input and output data (supervised learning see Sect. 5.4.1), clustering and association of non classified data (unsupervised learning see Sect. 5.4.2) as well as reinforcement learning, where an agent is in exchange with an environment as a source of interference (see Sect. 5.4.3). The method chosen depends entirely on the application case and should be carefully considered before starting to create a model.

5.6.1 Convolutional Neural Network (CNN)

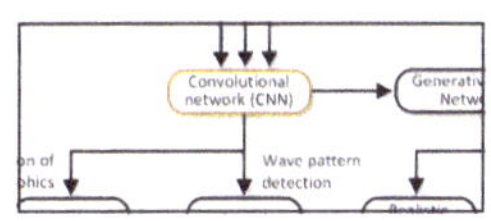

Convolutional Neural Networks (CNNs) belong to an extended form of networks that, similar to ANNs, are composed of neurons with learnable weights and a bias. The fundamental difference is the application of these networks: CNNs explicitly assume that the input consists of graphics. As a consequence, CNNs are composed of pure feedforward connections, (see Sect. 5.5.2). For the training of the weights, the backward propagation method is usually used (see Sect. 5.5.3).

Usual ANNs are unsuitable for scaling single neurons to a full pixel level of images. A simple graphic with 32X32 pixels and 3 RGB color codes per pixel would have 3072 connections (weights) in the hidden layer if an input neuron was completely connected by all combinations. If one assumes that a significantly higher number of hidden layers is desired to increase the performance of the model and that graphics can have much higher resolutions, the total number of weights would exceed the computational load of any computer. Hence, a completely linked connection of all neurons would prove to be inefficient.

CNNs divide the 3D pixel color code of a complex graphic into many small segments called cascades. From now on, a neural network is divided into subareas—each subarea represents a cascade and is individually trained. This highly efficient approach allows the number of connections to be reduced abruptly. This has the further advantage that cascades of certain image areas can train special image features, which, when all cascades are combined, provide more detailed information about a graphic.

Image Recognition

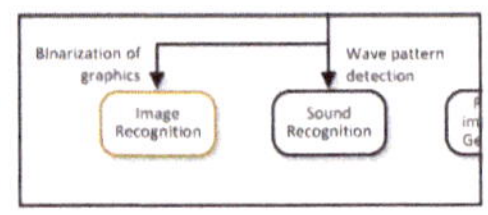

Image Recognition is an interdisciplinary field of science that focuses on training algorithms to examine digital images, videos or video sequences at a high level of detail. This includes methods for capturing, processing, analyzing and extracting multidimensional visual data from the real world. Image recognition can be interpreted as the decoding of symbolic information from image data, which in its inner core can be constructed of geometry, physics and statistics. In principle, the method of image recognition is comparable to a classification of the supervised learning process (see Sect. 5.4.1).

From an engineering perspective, this science is used to automate tasks that the human eye and brain cannot comprehend due to their complexity. A common problem of image processing is to determine whether image data contain specific objects, features or image sequences. For this purpose, the process of image recognition is divided into different cascades.

Using the example of image recognition, the transformation chain and the process of a CNN will be illustrated. For this purpose, an optical measurement of a 6-hole injector nozzle of a DI gasoline engine (snapshot of an injection pattern) at 270° crank angle before top dead center is fed into the network.

In Fig. 5.59, the first layer of the CNN recognizes the number of injection nozzles, the second layer deals with the injection angles, another layer with the geometric limitation by the combustion chamber walls, another layer focusing on the turbulence decay of the injection, etc. Furthermore, optical measurements can be post-processed by the application of filters, for example, by illumination, so that further layers can better interpret characteristic features of combustion such as temperature distribution, combustion speed, emissions and soot particles [63].

In the shown cascade, after feeding it with images, the hidden-layer process begins which consists of a convolution process and a pooling process, taking place in alternating order

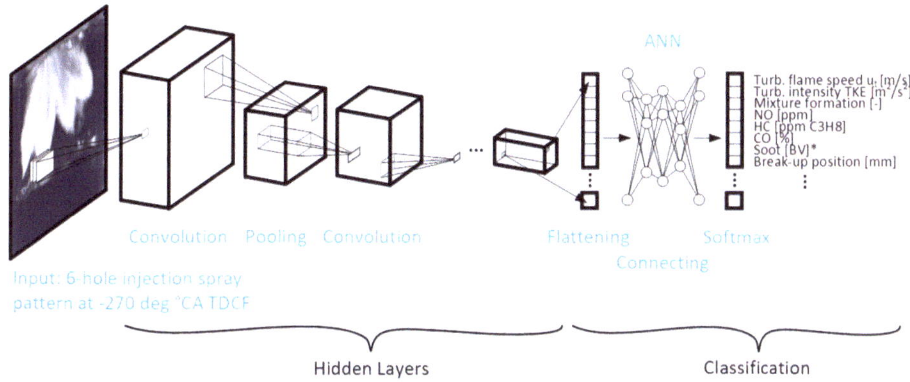

Fig. 5.59 Cascade structure of a CNN process[4]

to gradually break down desired features. Once this process is completed, the classification process follows. Here, the results of a matrix are transferred into 1D vectors and transformed into a softmax.

The Hidden-Layer Process

Convolution is one of the main building blocks of a CNN. The term derives from the mathematical combination of different sets of information to generate a function. By applying filters, so-called feature maps are generated. Each input node is subjected to matrix multiplication and the result is projected or summed up on the feature map. Especially for image processing, the matrices are composed of the 3D height, width and depth, and the contents correspond to the red-green-blue (RGB) color codes.

In the following, Fig. 5.60 shows the projection of a pixel-code area, which is projected through a convolution filter onto a result matrix.

In a hidden-layer process, the input is subjected to numerous convolutions, with each convolution using different filters. This results in different feature maps. Finally, all feature maps are combined as the sum of all convolution layers into an output matrix. As in all neural networks, activation functions are used to design nonlinear functions between inputs and outputs. A common activation function that is chosen is the ReLU function; see Fig. 5.32 in retrospect.

Since the size of a feature map becomes smaller than that of its input matrix due to a convolution, an intermediate step must be initiated to restore the original size. For this purpose, the padding method is used, and empty spots of the result matrix are filled up with zeros.

[4] BV=Blackening Number or smoke degree.

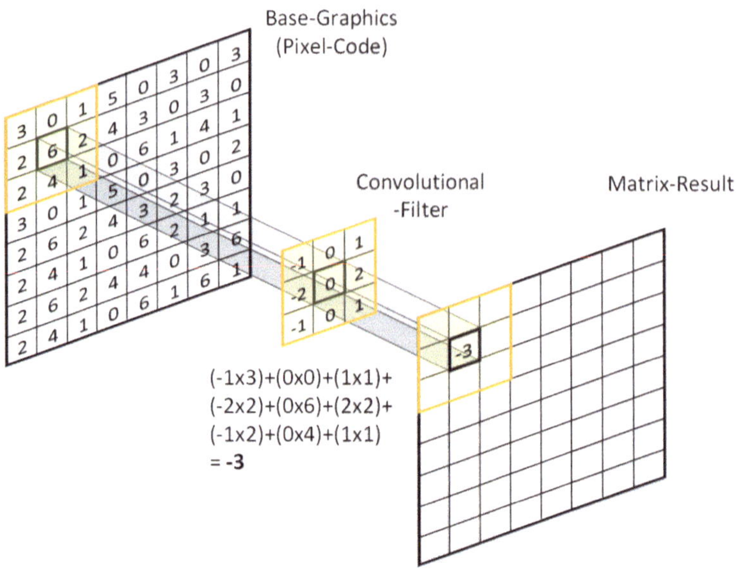

Fig. 5.60 Conversion of an RGB pixel code by a convolution filter (convolution process)

After convoluting and padding, the next process called pooling is performed. This has the function of smoothing the result matrix and removing possible irregularities or outliers. In addition, this process continuously reduces the dimension of the matrix, which reduces the computational load of the network. Different approaches for pooling are presented in the literature. Among the most important ones are Average Pooling (projection of the mean value), Max Pooling (projection of the maximum value) and RMS Pooling (projection of the quadratic mean value) as summarized here (Fig. 5.61).

The Classification Process

The main purpose of the classification process is to transform the results of the hidden-layer process into tangible quantities. According to Fig. 5.59, this section is divided into Flattening, Connection and Softmax.

Fig. 5.61 Pooling method for data filtering [64]

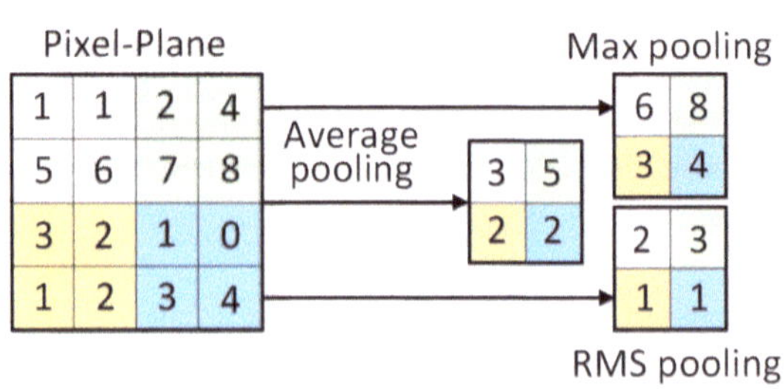

The Softmax function comes from mathematics and has the function of normalizing absolute numbers of different features and units. The normalization process projects all numbers onto a range of [0,1] so that they can be interpreted as probabilities. Consequently, the result of a CNN process is a statistical analysis serving a probability statement. It is important that through such an intensive learning process, the network is able to decipher independently which functions have to be optimally placed on which level. In this way, each layer learns to convert its input data into a more abstract and coordinated representation. If a statistical analysis is not desired, the value range of the output can also be transformed back to the original unit.

Image recognition has become a far-reaching term that combines a variety of technologies. One of the more prominent applications is the principle of **Identification**. Here, it is of interest to identify individual features of an object in order to compare them with other features from a database, for example. The identification by scanning of QR codes, eyes (iris scan), fingerprints or handwriting is one of the most prominent applications.

Object recognition or **object classification** belong to the methods that aim to identify single or multiple defined objects or object classes within an image. Blippar, Google Goggles and LikeThat are examples of independent software products in this segment.

A subcategory of this, called **Shape Recognition**, deals with the investigation of class features in the narrow sense. In contrast to object recognition, nested layers are analyzed and distinguished, such as head, shoulders and arms in an image of a person or animal, to which the averted form **Face recognition** also belongs.

Graphics Detection is a method that scans images to identify special features. It is mainly used to analyze parts of interesting image data and to conclude concrete interpretations regarding the content. In medical applications, for example, this includes the detection of possible mutated cells or abnormal tissue from high-resolution CT or MRI images.

Content-based image retrieval is based on a fast search algorithm that is able to check a huge pool of images (example: circulating Internet data) for specific content. The search criterion can be to search for images with similarity to a reference image or by entering any search criterion (example: search all images from which animals are found).

In addition to the categories mentioned above, there are a number of other application-related possibilities such as **Pose estimation**, for estimating the orientation of certain objects relative to the camera (example: supporting a robot arm when retrieving objects from a conveyor belt for industrial applications), and **Motion analysis**, for detecting moving images (video recordings) [65, 66].

The following examples will give some ideas on how image recognition techniques can be applied to powertrain development in a goal-oriented way.

Example 12: Influence of injection characteristics on combustion and emissions

The image recognition is suitable for breaking down a large number of graphics into their characteristics in order to draw conclusions about their content. One topic from the field of powertrain development that is ideally suited for this purpose is the evaluation and processing of optical measurements using high-resolution low- and high-speed cameras. In order to use the cascade of the presented example from Fig. 5.59, it is shown here to what extent large data sets of injection cone images can be used to train neural networks. Three different injectors with the following geometries are the basis for preceding optical measurements, see Fig. 5.62.

The training phase of a CNN is carried out with a total of 750 images. Each injector is subjected to a series of measurements, with 50 discrete crank angle images. The operating load is in the range of 0–40% of the maximum possible load. The following pictures show the result of a measurement series. On the left side, a standard image is shown, and on the right side a special color filter is applied to emphasize certain image features in order to simplify later recognition process (Fig. 5.63).

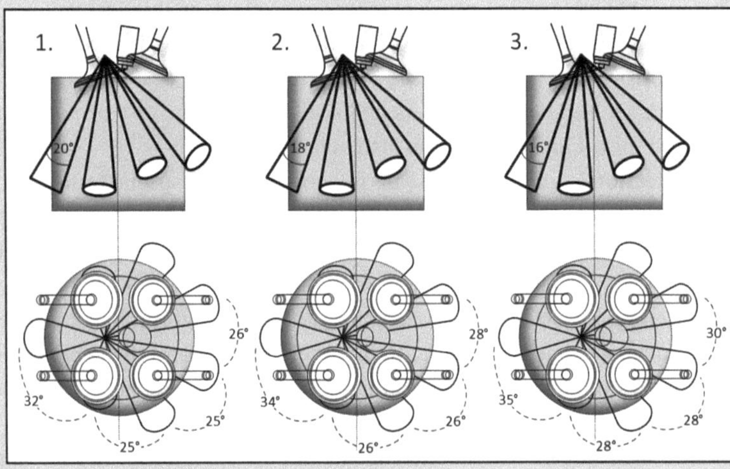

Fig. 5.62 Geometries of three different fuel injectors

The results of combustion from the resulting injection variations are shown here. Filters are also used here to enhance the visual characteristics, see Fig. 5.64.

Finally, for the training phase of the CNN, desired parameters such as turbulence, mixture and emissions which were determined by means of measuring devices, are put into relation with the images (Fig. 5.65).

Images not involved in the training are always used to validate the model. In principle, it is possible to reverse the procedure. In this case, the CNN would generate images of injection features and the corresponding combustion if the desired output variables are specified. The model is thus able to provide precise suggestions as to what the geometric characteristics of the injectors should look like in order to meet desired physical target values.

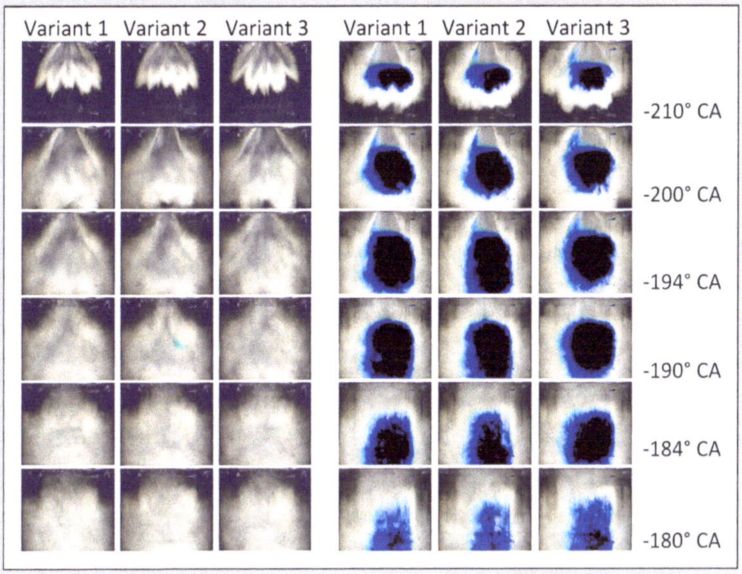

Fig. 5.63 Optical measurements of injection characteristics for different geometries

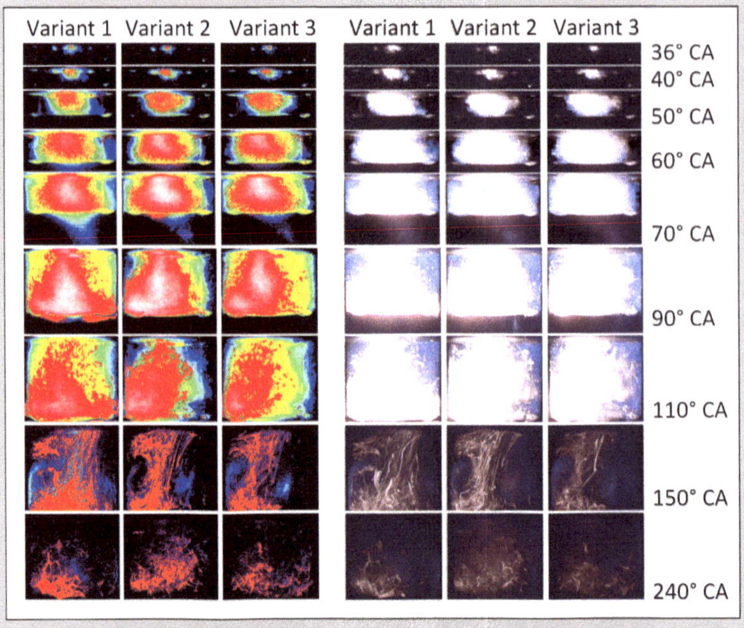

Fig. 5.64 Combustion optics: Application of fluorescence filters to enhance image features: diffusion flame (left) and soot formation (right)

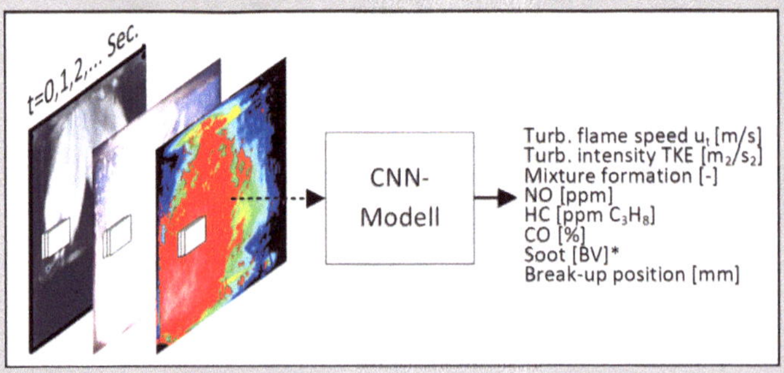

Fig. 5.65 CNN to predict specific quantities (turbulence, mixture formation, emission, etc.)

Example 13: Flow simulation

In this example, a CNN is developed to reflect 2D/3D flow calculations. The main focus is on the fact that it is in principle feasible to predict transient, i.e. time-dependent phenomena of Newtonian flows.

As long as a neural network is provided with the necessary inputs, it is in principle possible to implement shear rates proportional to shear stresses as well as the viscous behavior that obeys the equations of Navier-Stokes. If a high quality of agreement with classical 3D-CFD computation can be achieved, massive time and cost savings will result, particularly since the computational time reflects the bottleneck of development processes. In retrospect, Fig. 4.16 has qualitatively represented the savings of the calculation time that are possible with AI compared to conventional methods.

Using the example of a plate with variable dimensions in height, length and width, airflow is examined, which enters the plate at the entrance (top right) and exits on the bottom left; see Fig. 5.66. The state is changed by making a circular or elliptical incision in the plate. This incision is shifted on the coordinate system within the plate and, in the case of an ellipse, rotated randomly in order to train all possible effects of the deflection of the flow.

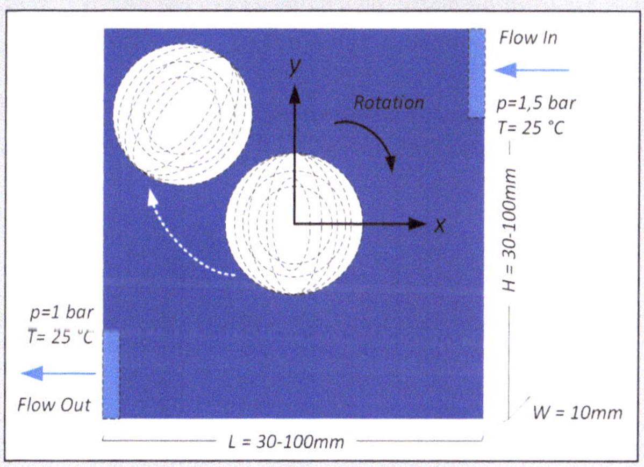

Fig. 5.66 Training phase of a flow through a plate with variable geometry (variable circular or elliptical incisions)

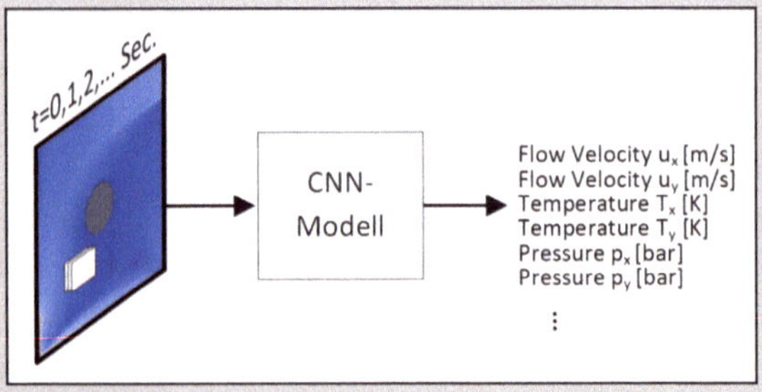

Fig. 5.67 CNN for generating time-discrete outputs of flow results

Then each variant is calculated by 3D-CFD simulation. The results of the calculation provide time-discrete snapshots, which are used for the training of the CNN (5000 images in total).

Figure 5.67 shows the principle structure of the model. With 3D-CFD, the CNN model is trained and time-discrete results for the velocity vector (u_x, u_y)*, the temperature vector* (T_x, T_y) *and the pressure vector* (p_x, p_y) *are determined. The 6D output vector* $(u_x, u_y, T_x, T_y, p_x, p_y)$ *reflects the flow state in each coordinate point.*

In contrast to the training phase, which consists of the generation of flow patterns of plates with a single incision, the validation phase also examines plates with two or more incisions. This helps to understand how flexible the CNN model is and how it is able to adapt to new and previously unknown planes. In the following, 3 different plates are presented for the validation, all with the dimensions $(x = y = 50mm, z = 10mm)$ (Fig. 5.68).

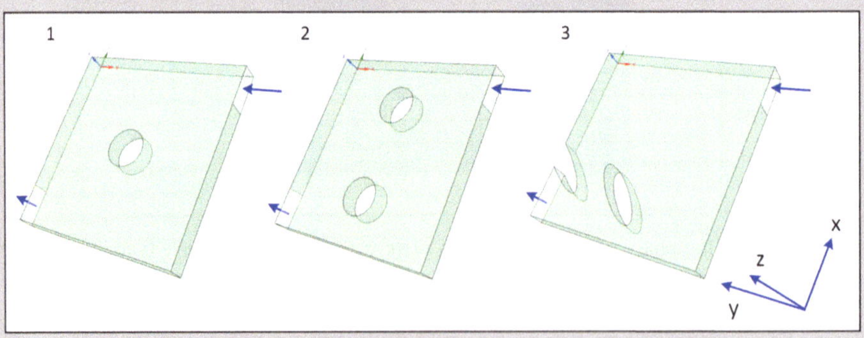

Fig. 5.68 Validation of the CNN model using flow-through plates with different incisions

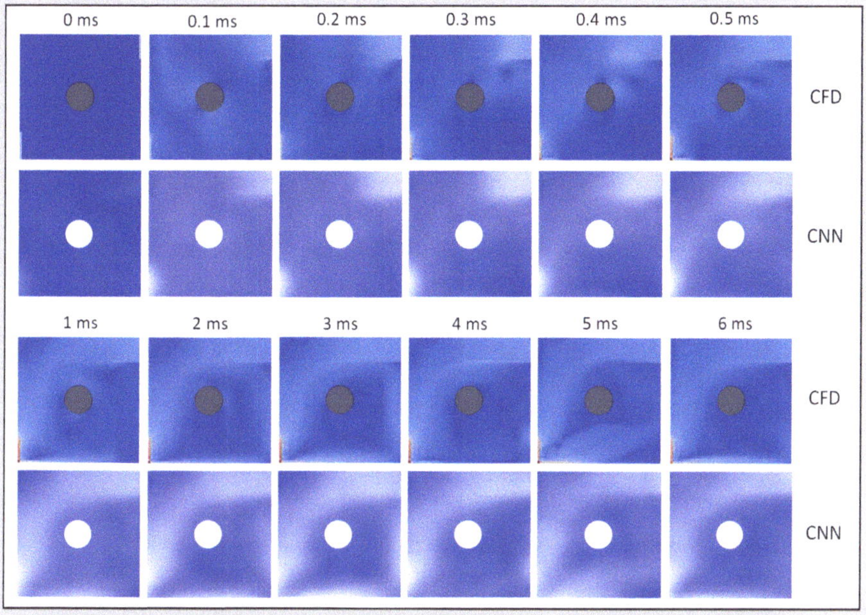

Fig. 5.69 Plate flow with central, circular incisions in comparison: CNN versus 3D-CFD

Figure 5.69 compares the result of the CNN model for plate variant 1 and a central hole incision with the 3D flow calculation (CFD). The CNN model generates transient flow images ranging 0ms–6ms. It is remarkable that it is able to reproduce recurring flow fluctuations (see 4ms–6ms) that occur in a phase before convergence is reached.

The example of plate variant 2 clearly shows that the interaction of the flow can be transferred to 2-hole incisions. Realistic images can also be generated here compared to that of CFD calculations (Fig. 5.70).

Finally, the validation on plate variant 3 shows that the flow can be qualitatively simulated even along complex and asymmetrical incisions (here elliptical shapes), one of which is adjacent to the plate edge. Here also, the effects of short-circuit currents, which are reflected in the snapshots ranging 0.5ms–3ms, are remarkable (Figs. 5.71).

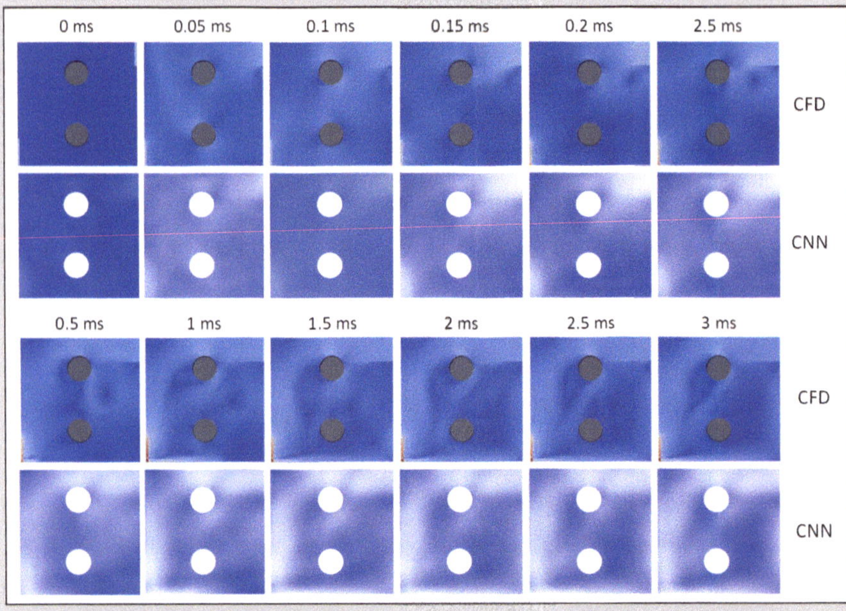

Fig. 5.70 Plate flow with two circular incisions on the transverse axis in comparison: CNN versus 3D-CFD

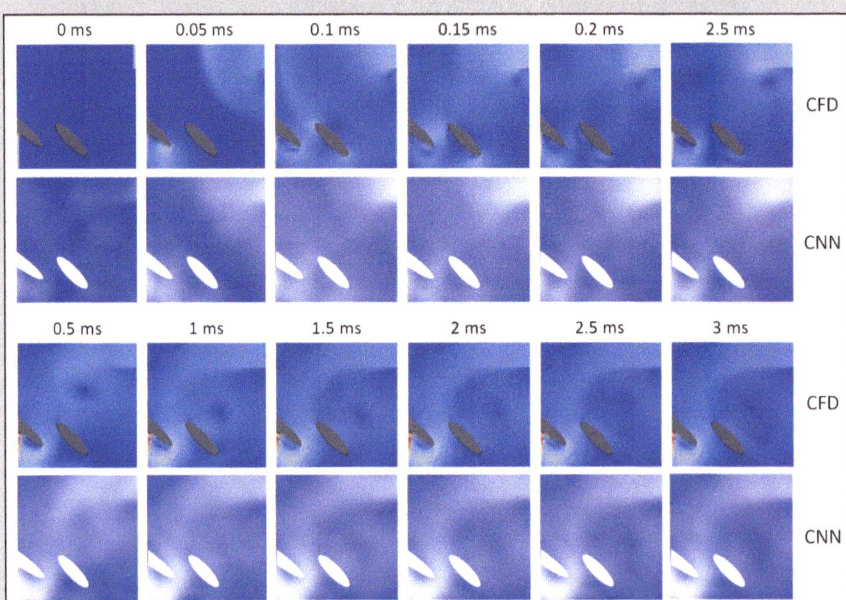

Fig. 5.71 Plate flow with asymmetrical incisions in comparison: CNN versus 3D-CFD

Sound Recognition

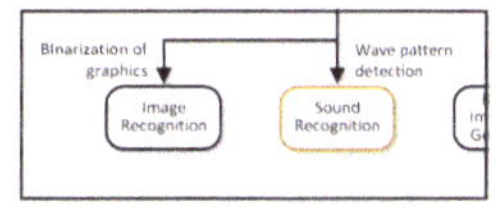

For over 120 million years, sound recognition has been one of the primitive survival instincts of all life forms. The sensitivity and performance of the sense of hearing can vary greatly. Speech and hearing together form the ability to communicate, and depending on lifestyle and living environment, the sense of hearing is evolutionarily adapted to enable the ability to communicate within individual species and to allow the auditory recognition of natural enemies.

The interaction of the ear and the brain as the central processing unit enables sound to be characterized for different features. The localization of sound combines the recognition of the **sound direction** and the **sound distance**. As an indication for the determination of the distance, the sound level or level differences of different sound sources relative to each other play an important role. The **frequency spectrum** of the sound provides the information about the sound source itself and can additionally be helpful for determining the distance to the source. As, for example, a distant sound is perceived as dull, a further characteristic is the **motion of a sound** that is perceived while one of the mentioned characteristics changes relative to the listener.

Sound recognition is a decade-old field of science that traditionally deals with pattern recognition theories. The underlying analysis methods use algorithms for data processing, feature extraction and classification of sound. More recently, the field has benefited enormously from the advances in artificial intelligence. Especially the data collection through the method of Big Data as a statistical analysis tool for real data has proven to be highly efficient. In combination with the deep learning network type (CNN), Sound Recognition is now evolving into a modern technology that aims to solve the processing of a sound according to human processing concepts. This is done by imitating neuronal and biochemical mechanisms underlying unconscious thinking. The power of modern computers helps to exceed the performance of a human brain by far.

Conventional method of sound recognition

Figure 5.72 shows a common sequence of different techniques for the sequential decoding and classification of sound features from the time domain [67].

Sound

A common sound recognition process starts with a sound being recorded, which is received by a microphone as a time signal. The higher the quality of the recording, the easier the recognition process can be designed. Background noises or room reverberation, for example, are

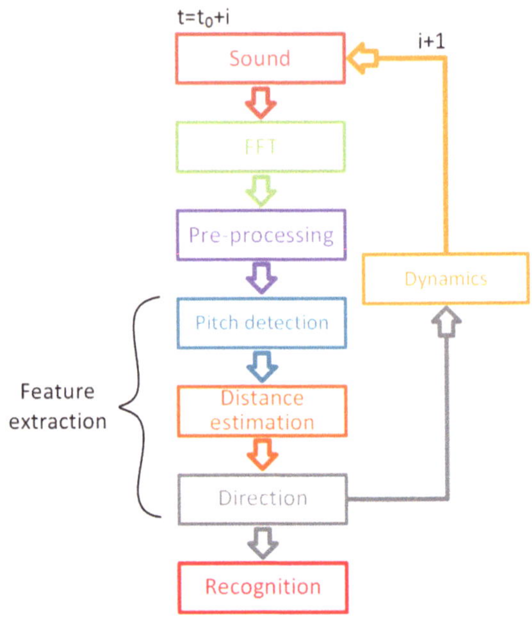

Fig. 5.72 Conventional Sound Recognition: Sequential extraction of individual sound features

unwanted effects that disturb the sound and make the later recognition process more difficult. The quality of the transduction unit in the microphone (conversion of a sound pressure into an electrical signal) can also have a great influence on the quality of the sound recording.

FFT
In a few cases, the information of a sound wave can be sufficient to extract features for sound recognition. However, the variability and the resulting complexity of sounds usually require the execution of a Fast Fourier Transformation (FFT) and the transformation of the wave signal into its spectral components. From this, all relevant frequency orders with the associated amplitudes and phase information are obtained, by which the characteristics of a sound signal are established; see retrospectively Sect. 3.5. The FFT is intended for periodic signals with a fixed periodic length. If a recorded sound is variable (dynamic), a fixed period length must be cut out of the sound signal.

Pre-processing
Pre-processing plays an important role in eliminating irrelevant sources on the recorded audio soundtrack, which later facilitates the accuracy of the recognition process. It includes the filtering of background noise, smoothing of dynamic noise, endpoint detection for periodic closure of a sound signal (for non-periodic signals), determination of a suitable window function and suppression of reverberation and echo.

Pitch Detection
Pitch detection is used to characterize dominant tones. This ensures that only frequencies with a high energy density compared to others in the overall spectrum are sorted out. The consideration of the cleaned spectrum can be important for the classification of a noise source.

Distance estimation
The distance between the microphone and the sound source, especially with moving sound sources, can be estimated by the sound pressure level. For this purpose, a relative calculation is made in relation to existing background noise.

Direction
If a sound source is moving, it may also be of interest to consider the direction of movement relative to the microphone. This can be realized if several microphones (at least 2) are mounted in opposite directions, so that the relative change of the different sound pressure levels can be taken as a reference.

Dynamics
In the process phase "FFT", the problem of the periodicity of a sound signal was mentioned as a prerequisite for a Fourier synthesis. If a sound is non-periodic, this problem can be solved in the last phase by stringing together many non-periodic signals of the FFT. The result is a quasi-continuous picture of a tone, which allows examining and evaluating even time-dependent signals.

Sound recognition with AI
Sound recognition based on neural networks corresponds in principle to a classification of the supervised learning method (see Sect. 5.4.1). While CNNs only allow images as inputs, sound recognition is performed using the same principles of image recognition. After an acoustic sound signal has been transformed into a spectrogram, i.e. into a graphical representation, using the FFT method, a CNN method can be applied. Fig. 5.73 shows exactly this process sequence [68].

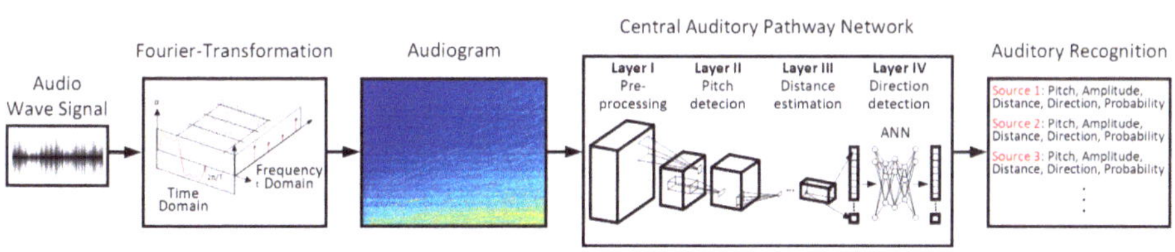

Fig. 5.73 Sound Recognition: Sequential extraction of individual sound features by CNN

Noise vibration and harshness (NVH), i.e. the acoustic data acquisition and analysis of structure-borne and airborne sound, is one of the major topics in the development of motor vehicles and offers far-reaching application potential for sound recognition. Some important applications are listed as follows:

- Body and chassis
  - NVH and structural dynamics for car body vibrations especially for lightweight components in the medium frequency range;
  - Eigenvalue analyses of brakes;
  - Rolling noise in the passenger compartment and for driving past depending on tire profile and wheel suspension;
  - Aeroacoustics: Audible vibrations in the vehicle interior due to external currents and sound pressure.
- Powertrain
  - Engine acoustics, torsional vibration dampers and injection systems;
  - Valve train;
  - Transmission and differential gear.
- Electric mobility and vehicle electronics
  - Electrical drive, battery and power electronics (PCU);
  - Electronic boards and control units (ECU) and GCU.

Voice Recognition has been used for many years in criminal investigations for the automated comparison of stored data. It is also used for identification monitoring, alarm systems and personal identification at banks. Acoustic oceanography and the science of animal noises and communication behavior have provided sound recognition with great insights over the last decade. More conventional in use are natural speech recognition systems like Siri and Alexa or music recognition applications like Shazam and Soundhound, which use the same techniques of artificial intelligence.

Example 14: Powertrain acoustics
Powertrain acoustics offers a wide range of applications for the sound recognition process. This is due to the fact that acoustic measurement data can be generated easily, but spectral analyses can only be differentiated at great expense. Depending on the measurement signal, the data can vary by different levels of uncertainty due to measurement noises. An acoustic engineer is therefore often faced with the task of breaking down measured sound signals, which are composed of a superposition of different noise sources, into their individual components. Along with extensive experience, components of special frequency spectrums can be assigned to a noise source—the task becomes more complex when destructive interference causes signal sources to be canceled out so that they cannot be recovered again.

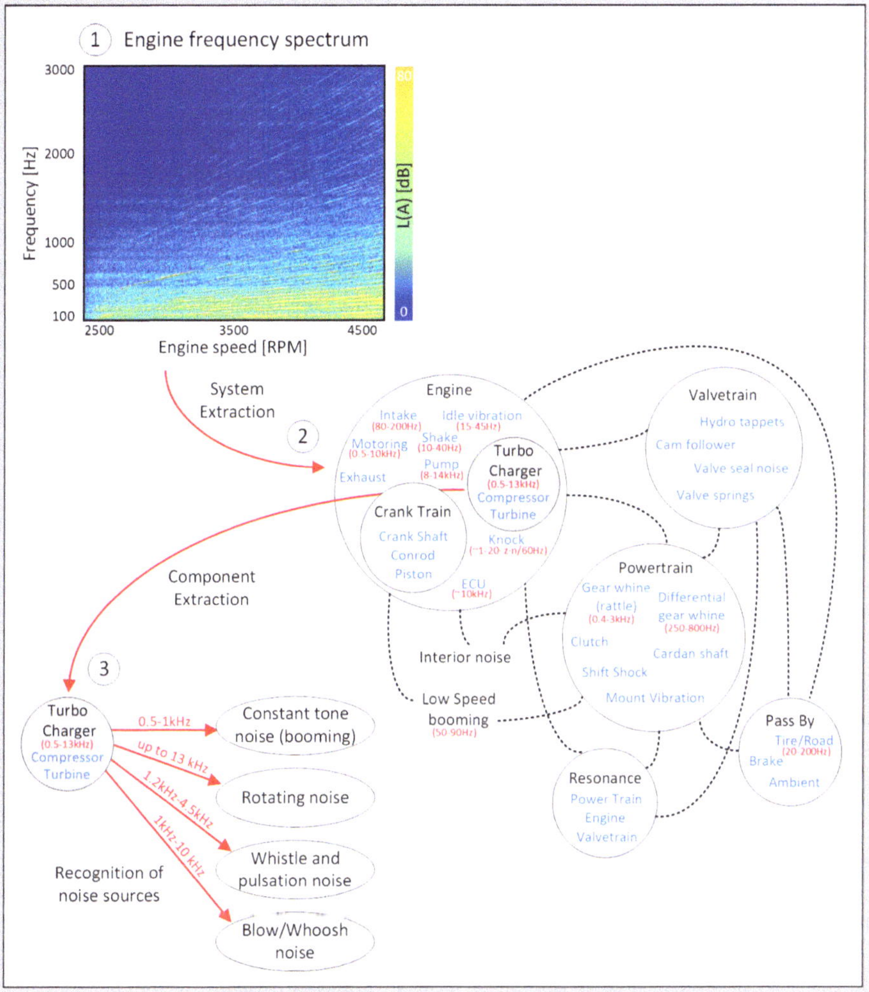

Fig. 5.74 Sound recognition applied to the frequency spectrum of an engine to recognize turbocharger specific noises

As can be seen in the next example, a CNN is fed by the frequency spectrum of a recorded sound that was measured on a motor. The model is able to separate all superimposed sounds from each other and assigns them to individual systems. Furthermore, it determines individual noise sources, as in the example of the turbocharger, and splits these into individual frequency ranges. The sound signal is then completely characterized, which enables the acoustic engineer to make a detailed pre-diagnosis of the existing system (Fig. 5.74).

5.6.2 Generative Adversarial Network (GAN)

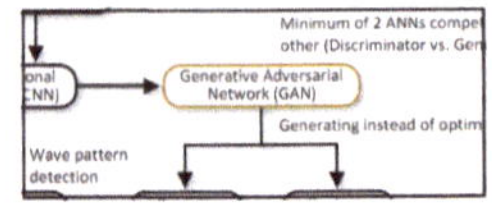

Generative Adversarial Networks (GAN) are part of a highly advanced group of algorithms developed in 2014 by a US computer scientist named Ian Goodfellow, which has established itself remarkably fast for artificial intelligence applications. The concept, which was originally based on the supervised learning method, has now also proven to be very powerful for unsupervised and reinforcement learning. The special feature of a GAN is that two neural networks are used simultaneously—a so-called generator and a discriminator. These networks compete with each other—while the generative network generates synthetic data, i.e. data that does not originate from reality, and real data is simultaneously fed into the system. The discriminator faces the challenge of evaluating the incoming data and checking it for authenticity. The interaction of these neural networks promises a highly efficient and autonomous training.

While the discriminator divides a given input into a cascade structure in order to perform a feature characterization on different levels, the way the generator works is contradictory. In each feature level, it is able to change properties by a reciprocal convolution. The following example shows the functioning of a generator and a discriminator by the example of microscopic fracture structures of a metal material (Fig. 5.75).

The GAN is interpreted as a competitive process because of the way it works. The training goal of the generative network is to increase the error rate of the discriminative network, i.e. to fool the discriminator by generating data that appear realistic so that the discriminator classifies them as non-synthesized.

If the discriminator is fed with a real and a synthetically generated image, it generates a function $D(x)$ as an output signal, which corresponds to a probability x that states whether the input corresponds to a real image or not. The goal is to maximize the function $D(x)$. To measure a representative result, the cross- entropy function is used.

$$\underset{D}{max}\, V(D) = E_{x_{t+1} \sim p_{data}}(x_{t+1})[log D(x_{t+1})] + E_{z \sim p_{noise(z)}}[log(1 - D(G(z)))] \quad (5.19)$$

The generator on the other hand also tries to generate images at a maximum possible value for $D(x)$.

$$\underset{G}{min}\, V(G) = E_{z \sim p_{noise(z)}}[\log(1 - D(G(z)))] \quad (5.20)$$

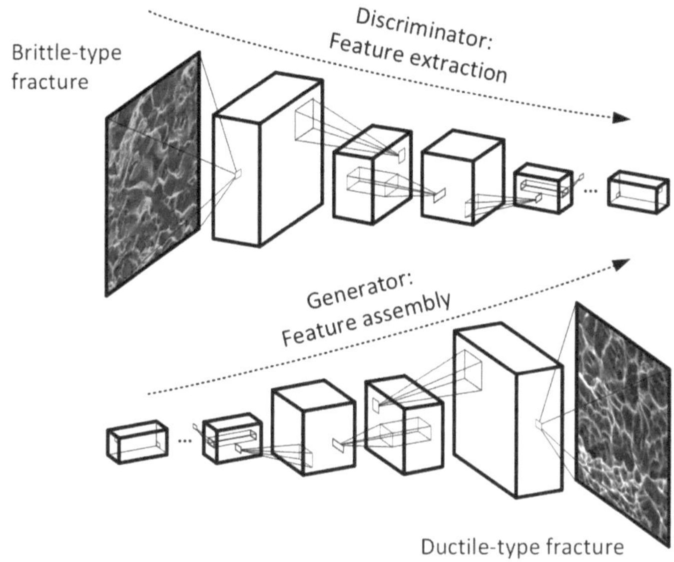

Fig. 5.75 Functioning of the discriminator (top) and the generator (bottom) at microscopic fracture structures of a metal material

In summary, a GAN is often interpreted as a MinMax game, where the generator G tries to minimize the function V and the discriminator D tries to maximize it (Fig. 5.76).

$$\underset{G}{min}\ \underset{D}{max}\ V(D, G) = E_{x_{t+1} \sim p_{data}}[\log D(x)] + E_{z \sim p_{noise(z)}}[\log(1 - D(G(z)))] \tag{5.21}$$

Monitoring and training are controlled by an agent. For a sample matrix (input matrix) $[x]$ with m dimensions $x_1, x_2, \ldots, x_m$ and a noise matrix $[z]$ with m dimensions $z_1, z_2, \ldots, z_m$, an agent transfers the following equation to the discriminator:

$$\nabla \Theta_D \frac{1}{m} \sum_{i=1}^{m} [log D(x^i) + log(1 - D(G(z^i)))] \tag{5.22}$$

The generator on the other hand is monitored with the following equation:

$$- \nabla \Theta_D \frac{1}{m} \sum_{i=1}^{m} [\log(1 - D(G(z^i)))] \tag{5.23}$$

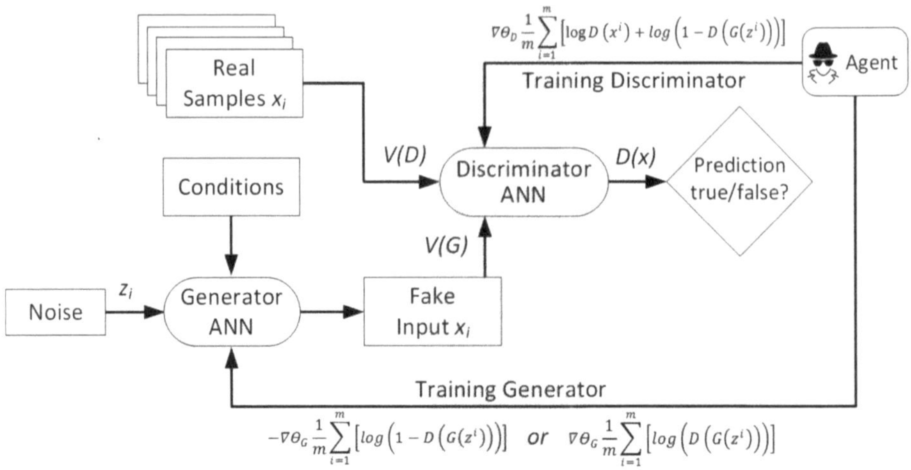

Fig. 5.76 Working process of a GAN

With the gradient descent method, one quickly encounters the problem that the discriminator defeats the generator early on due to its stronger function weighting, which means that in the early training stage non-real images can be quickly distinguished from real ones. To solve this problem, it is a good idea to program the GAN with an alternative function to retransmit the gradients [69]:

$$\nabla\Theta_D \frac{1}{m}\sum_{i=1}^{m}[\log(D(G(z^i)))] \tag{5.24}$$

GANs are currently still mainly used in the field of image processing and image generation. Among the best known applications are the realistic generation of photographs of human faces, manipulation of human poses, and "photo-aging" which is used by many cell phone applications. In the cartoon area, GANs are used to create new characters, and in the fashion design area to create suggestions for new patterns, designs and fresh inspirations. On the Internet, astonishing artworks are circulating that have been synthetically created with the help of existing works by famous artists and are amazingly similar to them. In the area of technical design, GANs are used in the modeling of 3D objects to generate new systems of objects and so to create fresh ideas, such as in the development of sports prostheses.

Example 15: Application of electronic control units

In the field of powertrain development, the basic idea of synthetic data generation can be transferred excellently to the base application of ECUs, the application of hybrid strategies or the power application of GCUs and PCUs. In principle, huge amounts of data are required to cover different stationary and transient operating ranges. For an internal combustion engine, a base application usually starts with the mapping of the air path for the steady state. For this purpose, the air mass flow is represented as a function of 1. intake valve timing, 2. exhaust valve timing, 3. boost pressure, 4. wastegate position, 5. throttle position and 6. engine speed. This is also done for different 7. ambient temperatures and 8. ambient pressures. The strategic procedure may differ depending on the manufacturer. Due to the 8D system shown here, a very rough subdivision of all parameters into 5 categories each already results in $5^8 = 390.625$ *combinations. Even modern space-filling methods (see retrospective* Sect. 4.3) *are easily overstrained by an enormous measuring and simulation effort when a thorough and error-free application is desired.*

These types of problems can be greatly tackled by the concept of GANs. It is able to provide enormous help in this respect by independently generating a massive amount of virtual data that imitate real data. This eliminates the need for test-based generation by test bench measurements or simulation-based data generation, which saves a lot of time. Fig. 5.77 *illustrates a flowchart for this.*

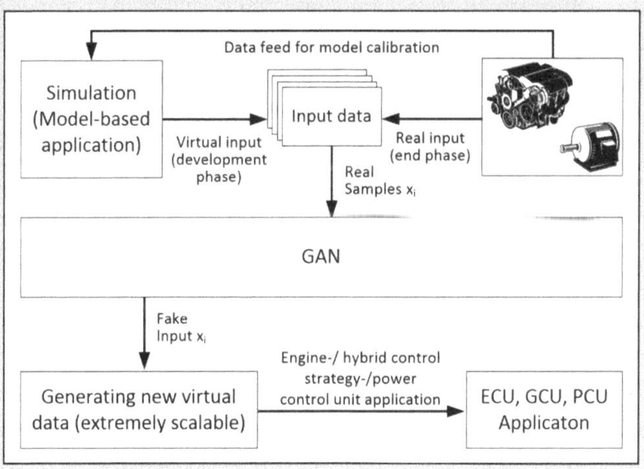

Fig. 5.77 GAN for an extremely scalable generation of data

5.7 Software

The science of data is a young and promising field that is developing rapidly, especially in the provision of data-based application software. This chapter intends to give an overview of ways for data handling and the underlying possibilities with regard to AI features, in the midst of numerous options. Which software and library package is the best one is left to the discretion of the respective application and may vary depending on it.

Many languages are now in use by data scientists and are very useful for individual applications. Among the most popular and widely used languages are Python, Scala and R, which have prevailed in the field of machine learning. Their now versatile and powerful libraries set them apart from individual solutions. Last but not the least, the constant development of additional package solutions provides the users a high degree of flexibility. Compared to commercially available simulation software tools for industrial applications, which were discussed in detail in Sect. 4, machine learning tools presented here are open-source solutions that can bring about significant cost savings in the value creation of industrial development processes.

For open-source software solutions, the source code is public and the copyright holder allows users to modify and adapt it for their own individual purposes. Because of the flexible applicability and the freedom that is offered, open-source solutions enjoy a very high acceptance.

Especially for students and other academic fields, open-source solutions offer free or low-cost access and a platform to an open community that is always in exchange to support each other. This accelerates the development process of the software automatically. Compared to commercial software, open-source solutions are considered safe to use. Possible errors in the code that were overlooked by the original developers of a program can be quickly detected and corrected under their own steam. By informing the community, an otherwise typical software update and, combined with a new release process is eliminated and hence, a substantial amount of time can be saved.

In terms of stability and long-term project planning, open-source software offer further significant advantages over proprietary software. As a rule, they are based on open standards to ensure flexible integration. Even after development projects have died, tools are available to users for an indefinite period of time.

In addition to the idea of a community, values and principles such as sustainability, efficiency and cost savings are increasingly becoming the focus of digital corporate structures. Combined with the idea of lean management and the associated low-process and transparent development structures, open-source solutions are a revolutionary era. Not only as individual solutions for academic institutions and small companies as before, but especially for medium-sized and large companies, they are becoming more popular than ever.

Python

Python is a language for very general applications with a large number of stored and elaborated libraries. It is suitable for computer scientists, mathematicians and especially for engineers. It can be used for beginners and advanced machine learning applications as well as data analysis, operation and visualization.

Scala

Scala offers ideal solutions in dealing with Big Data. The combination of Scala and Spark allows the computing power of computers or graphic cards to be used optimally through cluster computing. This balances the high computing load especially with huge amounts of data, which is unique for Scala. The language has many powerful libraries for machine learning with close links to engineering. In the area of data analysis and visualization possibilities, Scala has a disadvantage compared to its competitors.

R

R was initially developed for statistical calculations and analyses and for historical reasons offers a range of high-quality packages for statistical data acquisition and visualization. Due to its strong interface and formatting capabilities to different syntax, R is especially well suited for research purposes due to this flexibility. Compared to its competitors, R is currently less suitable for applied purposes such as engineering.

Each of these languages is more or less suitable for certain types of tasks. The choice of a preferred programming language can be very subjective, so the developer should always choose the tool that is most convenient for him.

After the most suitable programming language has been selected, a further decision has to be made on the selection among available **Machine Learning** libraries. Depending on which AI application is preferred (Natural Language Processing, Search Tree, Supervised Learning, Unsupervised Learning, Reinforcement Learning, Deep Learning, etc. (see Fig. 5.5), a large number of different libraries are available. With the help of **visualization** packages, data operations can be post-processed and results visualized. Depending on the visualization option, this can lead to a better understanding and interpretation of the corresponding data. Libraries for **mathematics and engineering sciences** offer the possibility to make numerical data manageable and to perform complex mathematical operations and scientific calculations. These packages are also used to handle data that is difficult to interpret, such as text content. As the main component of data science, the field of **data operation and analysis** provides libraries that can be used to collect, clean and pre-processed data for subsequent analysis. Finally, packages for **research** support the idea of combining variable formats from different sources of syntax in order to provide the user the greatest possible freedom in handling (Table 5.3).

The following figure lists 20 popular libraries for each of the programming languages presented. This can help to give an overview in which category libraries can be assigned and which individual strengths and weaknesses have to be taken into account when searching for the best individual solution. The diagram shows only a brief outline among countless libraries that are available on the market and are being expanded with continuous growth [70].

Tab. 5.3 Open-source software and examples of some libraries for the application of AI

python

Library	Description
TensorFlow	• Framework for deep and machine learning • Provides modeling of ANNs for multiple data sets, main applications: object identification, speech recognition • Several layer-helpers come on top: tflearn, tf-slim, skflow etc., library is quick in new releases
learn	• Machine learning library based on NumPy and SciPy • It provides algorithms for many standard machine learning and data mining tasks such as clustering, regression, classification, dimensionality reduction etc.
PYTORCH	• Large framework allowing to perform tensor computations with GPU acceleration, dynamic computational graphs and automatic, calculate gradients • Offers a rich API, solving applications for neural networks • A library based on Torch, which is an open source deep learning library implemented in C with a wrapper in Lua
Keras	• High-level library for working with neural networks, running on top of TensorFlow and Theano • It is possible to use CNTK and MxNet as the backends
XGBoost LightGBM CatBoost	• Special libraries for optimized, scalable and fast implem-entations of gradient boosting to refine models by decision trees • XGBoost, LightGBM, and CatBoost are competitors that solve a common problem
dist-keras elephas spark-deeplearning	• Processing big amount of data with the use of distributed computing systems like Apache Spark expands the possibilities for deep learning • Dist-keras, elephas, and spark-deep-learning allow to train neural networks based on the Keras library directly with the help of Apache Spark
ELI5	• Often predictions of a machine learning model are not entirely clear, a challenge that eli5 library helps on • A package visualizing/debugging models, tracking the work of an algorithm. It supports for scikit-learn, XGBoost, LightGBM, lightning, sklearn-crfsuite libraries.
SciPy	• A core library for scientific computing based on NumPy • The data structure is a multidimensional array, implem-ented by Numpy. Package contains tools that help solving linear algebra, probability theory, integral calculus etc.
NumPy	• Provides libraries for scientific applications • Intended for processing large multidimensional arrays and matrices, with extensive collection of mathematical functions and implemented methods
python	• Natural language toolkit (NLTK) provides a set of libraries for language processing. With its help, text can be analyzed and processed in a variety of ways
statsmodels	• Provides many possibilites for statistical data analysis, such as statistical model estimation, tests etc. • Many machine learning methods can be implemented
spaCy	• A language processing library with excellent examples, API documentation, and demo applications • It supports almost 30 languages, easy deep learning integration with high accuracy and rubustness
gensim	• A library for robust semantic analysis, topic modeling and vector-space modeling, built upon Numpy and Scipy • Implementation of popular NLP algorithms, such as word2vec. Although having its own models, fasttext library can be used for learning of word representations
pandas	• A library providing high-level data structures and a vast variety of tools for analysis • Translates complex operations into simple commands • It contains many built-in methods for grouping, filtering, and data combining, as well as a time-series functionality
Scrapy	• A library used to create spider bots to scan websites and collect data. In addition, Scrapy can extract data from the API • The library is very handy due to its extensibility and portability
matplotlib	• A low-level library to create two-dimensional diagrams and graphs • Diverse charts, from histograms and scatterplots to non-Cartesian coordinates can be built
bokeh	• Bokeh library creates interactive and scalable visualizations using JavaScript widgets • Interactive abilities like a rotation of categorical tick labels, small zoom tool and customized tooltip fields enhancements.
plotly	• A popular library allowing to build sophisticated graphics easily. The package is adapted to work in interactive web applications. • Support for multiple linked views, animation, and crosstalk integration
Seaborn	• An higher-level API based on matplotlib library. It contains more suitable settings for processing charts • Providing a rich gallery of visualizations including time series, jointplots, and violin diagrams
pydot	• A library for generating complex- and non-oriented graphs. It is an interface to Graphviz, written in Python. • Strong in visualizing the structure of graphs, of neural networks and decision trees based algorithms

Scala

Library	Description
DEEPLEARNING4J	• A domain-specific language to configure deep neural networks • Implementations of restricted Boltzmann machine, deep belief net, deep autoencoder, stacked denoising autoencoder, recursive neural tensor network etc.
Spark MLlib	• Built on top of Spark, MLlib library provides machine learning algorithms, functional API for Java, Python, and R, but options for Scala are more flexible • MLlib is a library with core machine learning algorithms for classification, clustering, unsupervised learning etc.
IO	• A machine learning server for build and deploy predictive engines, built on Apache Spark, MLlib, and Hbase • Build, evaluate and deploy engines, implement individual machine learning models, and incorporate them into an engine
bigDL	• A distributed deep learning library for Apache Spark • Users can write deep learning applications as standard Spark programs, which can run on top of existing Spark or Hadoop clusters
summingbird	• A domain-specific data processing framework which allows integration of batch and online MapReduce computations and hybrid batch/online processing mode • It consumes and generates streams and snapshots, providing platform implementations for Storm, and Scalding
DeepLearning	• DeepLearning.scala is an alternative machine learning toolkit providing efficient solutions for deep learning • It creates complex dynamic neural networks through a combination of object-oriented and functional programming
Spark + H_2O SPARKLING WATER	• Combines machine learning algorithms of H2O with the capabilities of Spark • Integrating these two environments provides for users feed the results into H2O to build a model, make predictions, post-process results in Spark
Smile	• Statistical Machine Intelligence and Learning Engine (Smile), is similar to Python's scikit-learn. It is developed in Java and offers an API for Scala • The library provides applications and a large set of machine learning algorithms for classification, regression, nearest neighbor search, feature selection, etc.
Conjecture DATA SOLUTIONS	• Brings machine learning and data science to the challenges faced in the engineering industry • Conjecture System Identification helps engineers create simple, best-fit trends of their industrial process data, to better control, and diagnose their systems
akka	• Akka is a concurrent framework for building distributed applications on a JVM
kafka	• Apache Kafka is an stream-processing software platform written in Scala and Java. I provides unified, high-throughput, low-latency platform for handling real-time data feeds. It can connect to external systems via Kafka Connect and provides Kafka Streams, a Java stream processing library
Breeze	• Computing library for Scala carrying ideas from MATLAB's data structures and NumPy classes for Python • Provides data array manipulations, enables implementation Matrix/vector operations, probability/statistic functions, optimization, linear algebra, signal processing operations, plotting possibilities
Epic	• A natural language processing library as a part of ScalaNLP • Known as a prediction framework which employs structured prediction for building complex systems
Puck	• A natural language processing librariey as a part of ScalaNLP • Mostly used as text parsers, with Puck being more convenient if one needs to parse thousands of sentences due to its high-speed and GPU usage
Saddle	• A data manipulation toolkit based on the „Frame" structure 2D indexed matrix • In total, there are five major data structures: Vec (1D vector), Mat (2D matrix), Series (1D indexed matrix), Frame (2D indexed matrix), Index (hashmap-like)
Scalalab	• A Scala's interpretation of MATLAB computing functionality. Ability to directly call the results of MATLAB scripts • It uses its own domain-specific language called ScalaSci. It gets access to scientific Java and Scala libraries • Most of the techniques are similar to Breeze and Saddle
Slick	• Scala Language-Integrated Connection Kit (Slick) is a library for creating and executing database queries offering a variety of supported databases H2, MySQL, PostgreSQL • To build queries, Slick provides a powerful DSL it supports both simple SQL queries, and strongly-typed joins of several tables
ADAM	• ADAM defines datasets and metadata standards that support: efficient generation, replication, and review of clinical trial statistical analyses, and traceability among analysis results, analysis data, and data represented in the Study Data Tabulation Model (SDTM)
VEGAS	• A visualization tool much more functional than Breeze-viz allowing plotting specifications such as filtering, transformation, and aggregation. It is similar in structure to Python's Bokeh and Plotly
Breeze-viz	• A plotting library developed by Breeze for Scala, based on Java charting library JFreeChart and has a MATLAB-like syntax • Although having fewer opportunities than MATLAB, matplotlib in Python, or R, it is very helpful in the process of developing and establishing new models

R

Library	Description
H_2O.ai	• H2O is an in-memory platform for machine learning that is reshaping how people apply math and predictive analytics to their business problems
mlr	• Mlr is an extensible framework for classification, regression, survival analysis, and clustering • Easy extension mechanism through S3 inheritance
dmlc XGBoost	• Implementation of the gradient boosted decision trees algorithm • Reach tools for regression, classification, and ranking problems • High speed and performance
caret	• Caret provides models for classification and regression applications • It serves as a powerful tool providing algorithms for creating predictive models
Microsoft LightGBM	• Light GBM represents generalized boosted regression models • Includes plenty of regression methods • Avalable tools for modeling variable selection and final stage precision
Prophet	• High-quality forecasts for time series data • Manages data that has multiple seasonality with linear or non-linear growth • Robust to missing data, shifts in the trend, and large outliers
dplyr	• dplyr is package useful for data manipulation, providing a consistent set of functions that help to solve the most common data manipulation challenges and it is designed to abstract over how a set of data is stored
	• Data.table is a package for fast aggregation of large data, using laconic flexible syntax and wide suite of useful functions • Friendly file reader and parallel file writer
lubridate	• Lubridate is a package providing a set of functions to work with date and time format data • Easy and fast parsing of date-time data • Expanded mathematical operations on time data
jsonlite	• Jsonlite allows a robust and quick parsing of JSON objects in R • Great tool for interacting with web APIs and building pipelines • Functions to stream, validate, and prettify JSON data
knitr	• Knitr is a transparent engine for easy dynamic report generaton in R • Enables integration of R code into LaTeX, LyX, HTML, Markdown, AsciiDoc, and reStructuredText documents
rmarkdown	• Markdown is a lightweight markup language with plain-text-formatting syntax. Its design allows it to be converted to many output formats • It has the abilitey to define new formats for custom publishing requirements
Slidify	• Generates reproducible HTML slides from R markdown • Allows embedded code chunks and mathematical formulas • Rich sharing and customizing ooportunities
ggplot2	• A data visualization package for the statistical programming which breaks up graphs into semantic components such as scales and layers
plotly	• Plotly provides online graphing, analytics, and statistics tools for individuals and collaboration, as well as scientific graphing libraries • Rich features and plenty of avalable charts • Abilities to make ggplot2 graphics interactive
ggvis	• Declaratively describes data graphics with a syntax similar in spirit to ggplot2 • Creates rich interactive graphics that can be played with locally in Rstudio or in the browser • Incorporates shiny reacitve programming model and dplyr grammar of data transformation
DataTables	• Displays R matrices and data frames as an interactive HTML table • Creates sortable tables with a minimum of code • Provides many useful features and styling options for tables
rCharts	• rCharts is a package to create, customize and publish interactive javascript visualizations from R using a familiar lattice style plotting interface
corrplot	• A package for graphical display of correlation matrices and confidence intervals • It also contains algorithms to do matrix reordering
lattice	• Lattice brings the proven design of Trellis graphics to R, considerably expanding its capabilities in the process • Lattice is a powerful data visualization system that is sufficient for most everyday graphics needs

Machine Learning ⬭ Maths & Engineering ⬭ Data operation & Analysis ⬭ Research ⬭ Visualization ⬭

Correction to: Powertrain Development with Artificial Intelligence

Correction to: Chapter 5 in: A. Mirfendreski, *Powertrain Development with Artificial Intelligence*, https://doi.org/10.1007/978-3-662-63863-7_5

Figures 5.28, 5.40 and 5.41 were replaced for legal reasons.

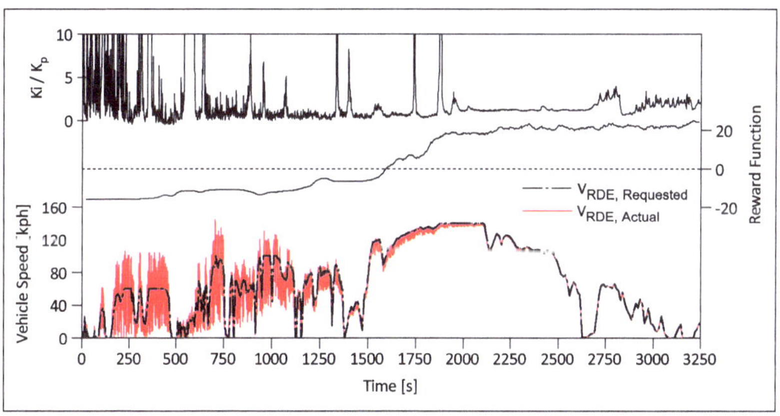

Fig. 5.28 Adaptive RDE speed control

The updated version of this chapter can be found at
https://doi.org/10.1007/978-3-662-63863-7_5

A. Mirfendreski, *Powertrain Development with Artificial Intelligence*,
https://doi.org/10.1007/978-3-662-63863-7_6

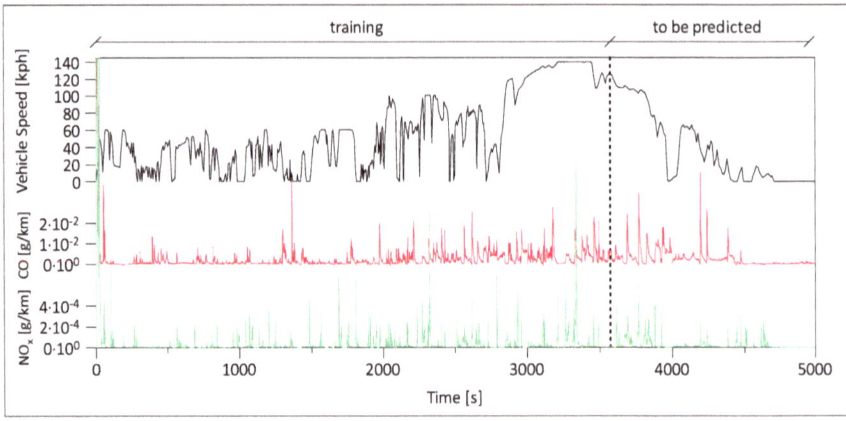

Fig. 5.40 Training and prediction of CO, NO_x emissions of the RDE test procedure

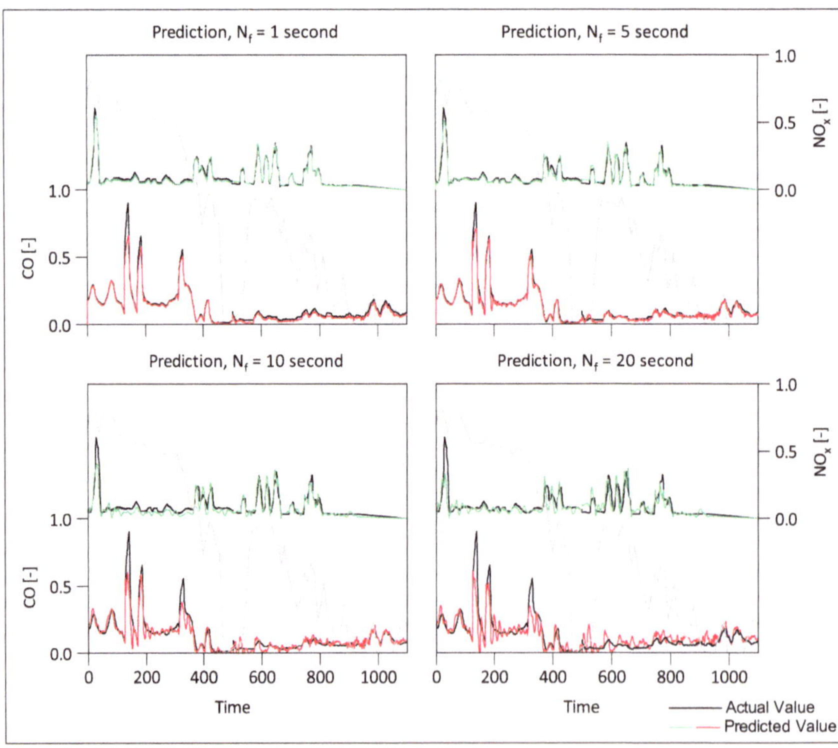

Fig. 5.41 Prediction of CO and NO_x in RDE for different future times steps N_f

Bibliography

1. R. Liedtke, *Die Industrielle Revolution* (UTB. Böhlau, Köln, Weimar, Wien, 2012). 978-3825233501
2. E.J.B.D.S. Montague, *Schöne Alte Automobile* (Gondrom, Bayreuth, 1978)
3. G. Goldbeck, *Gebändigte Kraft: Die Geschichte Der Erfindung Des Otto-Motors* (Moos, München, 1965)
4. J.H. Ferziger, M. Perić, *Numerische Strömungsmechanik* (Springer-Verlag, Berlin, Heidelberg, 2008). 978-3-540-68228-8
5. G.P. Merker, *Grundlagen Verbrennungsmotoren: Funktionsweise, Simulation*, 7th edn. (Messtechnik. Springer Vieweg, Wiesbaden, 2014). 978-3-658-03195-4
6. Gamma Technologies, editor. *GT Suite Flow Theory Manual Version 7.5*. Westmont, USA, 2015
7. L. Guzzella, C.H. Onder, *Introduction to Modeling and Control of Internal Combustion Engine Systems* (Springer-Verlag, Berlin, Heidelberg, 2010). 978-3-642-10775-7
8. M. J. W. Schouten, H. J. van Leeuwen. *Die Elastohydrodynamik: Geschichte und Neuentwicklungen*. volume 1207, Fulda, 1995. Eindhoven University of Technology, VDI Berichte
9. T. Mang, *Encyclopedia of Lubricants and Lubrication* (Springer-Verlag, Berlin, Heidelberg, 2014). 978-3-642-22647-2
10. C.H. Venner, A.A. Lubrecht, *Multi-Level Methods in Lubrication* (Tribology and Interface Engineering. Elsevier, Burlington, 2000). 9780080537092
11. E. Justi, *Spezifische Wärme Enthalpie* (Entropie und Dissoziation technischer Gase. Springer-Verlag, Berlin, Heidelberg, 1938). 978-3-642-99333-6
12. F. Zacharias. *Analytische Darstellung der thermodynamischen Eigenschaften von Verbrennungsgasen*. PhD thesis, Technische Universität Berlin, Berlin, 1966
13. T. Allison. *JANAF Thermochemical Tables, NIST Standard Reference Database 13*. National Institute of Standards and Technology. 10.18434/T42S31
14. G. Woschni, Die Berechnung der Wandverluste und der thermischen Belastung der Bauteile von Dieselmotoren. MTZ - Motortechnische Zeitschrift **31**, 491–499 (1970)
15. G. Hohenberg, *Experimentelle Erfassung Der Wandwärme in Kolbenmotoren* (Technische Universität, Graz, Habilitationsschrift, 1980)
16. K. Huber. *Der Wärmeübergang schnelllaufender, direkteinspritzender Dieselmotoren*. Dissertation, Technische Universität, München, 1990

A. Mirfendreski, *Powertrain Development with Artificial Intelligence*,
https://doi.org/10.1007/978-3-662-63863-7

17. M. Bargende. *Ein Gleichungsansatz zur Berechnung der instationären Wandwärmeverluste im Hochdruckteil von Ottomotoren*. Dissertation, Technische Hochschule, Darmstadt, 1991
18. H.G. Hahn, *Elastizitätstheorie: Grundlagen Der Linearen Theorie Und Anwendungen Auf Eindimensionale, Ebene Und Räumliche Probleme* (Vieweg + Teubner Verlag, Wiesbaden, 1985). 978-3-663-09894-2
19. F. Ziegler, *Technische Mechanik Der Festen Und Flüssigen Körper* (Springer-Verlag, Wien, 1998). 978-3-211-83193-9
20. G. Mueller, M. Möser, *Numerische Methoden Der Technischen Akustik* (Fachwissen Technische Akustik. Springer-Verlag, Berlin, Heidelberg, 2017). 978-3-662-55409-8
21. M. Bossert. *Einführung in die Nachrichtentechnik*. Technik 10-2012. Oldenbourg-Verlag, München, 2012. ISBN 9783486708806
22. J.W. Cooley, J. Tukey, *An Algorithm for the Machine Calculation of Complex Fourier Series*, Bell telephone system technical publications, vol. 4990. (Bell Telephone Laboratories, New York, 1965)
23. T. Butz. *Fouriertransformation für Fußgänger*. Springer-Verlag, Wiesbaden, 7. aktualisierte Aufl., 2011. ISBN 978-3-8348-8295-0
24. P. Jochem, *Alternative Antriebskonzepte Bei Sich Wandelnden Mobilitätsstilen* (KIT Scientific Publishing, Karlsruhe, 2012). 978-3-86644-944-2
25. International Council On Clean Transportation (ICCT). *CO_2 Emission Standards for Passenger Cars and Light-Commercial Vehicles in the European Union: Policy update*, 2019. URL www.theicct.org
26. J. Brokate, E. D. Özdemir, U. Kugler. *Der Pkw-Markt bis 2040: Was das Auto von morgen antreibt: Szenario-Analyse im Auftrag des Mineralölwirtschaftsverbandes*. Deutsches Zentrum für Luft- und Raumfahrt e.V., 2013
27. J. Adolf, C. Balzer, A. Joedicke, U. Schabla, K. Wilbrand, *Shell PKW-Szenarien Bis 2040: Fakten* (Trends und Perspektiven für Auto-Mobilität. Shell Deutschland Oil GmbH, Hamburg, 2014)
28. F. Bergk, K. Biemann, C. Heidt, W. Knörr, U. Lambrecht, T. Schmidt. *Klimaschutzbeitrag des Verkehrs bis 2050*. Umwelt Bundesamt, Heidelberg, edition 56, 2016
29. P. Hofmann, *Hybridfahrzeuge: Ein Alternatives Antriebskonzept Für Die Zukunft*, 2nd edn. (Springer-Verlag, Wien, 2014). 978-3-7091-1780-4
30. A. Ovens. *Bekanntmachungen zur Fahrzeugsystematik: Übersicht der Bekanntmachungen*. Kraftfahrt-Bundesamt, 2019. URL https://www.kba.de/DE/Statistik/Bekanntmachungen
31. *Infographics & Animations: Extracting business value from the 4 V's of big data*. IBM and Gartner, Stamford (Connecticut), USA, 2016. URL https://www.ibmbigdatahub.com/infographic
32. Q. Xiao. *Constructions and Applications of Space-Filling Designs*. PhD thesis, University of California, Los Angeles, 2017
33. S. D. Arnott, P. A. Lindsay. *Reducing Uncertainty in Systems Engineering through Defence Experimentation*. Researchgate GmbH, 2015
34. E. Aronson, T. Wilson, R. M. Akert. *Sozialpsychologie*. Psychologie. Pearson Studium, München, 6., aktualisierte Aufl., 2010. ISBN 3-8273-7084-1
35. A. Newell, J.C. Shaw, H.A. Simon, *Report On A General Problem-Solving Program* (Pittsburgh, Pennsylvania, USA, 1958)
36. F. Eisenführ, M. Weber, T. Langer. *Rationales Entscheiden*. Springer-Lehrbuch. Springer-Verlag, Berlin, Heidelberg, 5., überarb. und erw. Aufl., 2010. ISBN 978-3642028489
37. P. Kuhlmann, *Künstliche Intelligenz: Einführung in Machine Learning, Deep Learing, Neuronale Netze*, 1st edn. (Robotik und Co., Hannover, 2018). 9781983196065
38. V. Vowinkel. *Kommt die technologische Singularität?* HP Humanistischer Pressedienst, 2016. URL https://hpd.de/artikel/kommt-technologische-singularitaet-13480

39. R. Freshman. *Technological Singularity and A.I.: What is Technological Singularity?* Blog at WordPress.com, 2016. URL https://robertsfreshmanphysics.wordpress.com
40. I. Seifert, M. Bürger, L. Wangler, S. Christmann-Budian, M. Rohde, P. Gabriel, G. Zinke, *Potenziale Der Künstlichen Intelligenz Im Produzierenden Gewerbe in Deutschland: Studie Im Auftrag Des Bundesministeriums Für Wirtschaft Und Energie (BMWi) Im Rahmen Der Begleitforschung Zum Technologieprogramm* (Begleitforschung PAiCE, Berlin, 2018)
41. P. Hart, N. Nilsson, B. Raphael, A Formal Basis for the Heuristic Determination of Minimum Cost Paths. IEEE Transactions on Systems Science and Cybernetics **4** (2), 100–107 (1968). https://doi.org/10.1109/TSSC.1968.300136
42. H. Braun, *Neuronale Netze: Optimierung Durch Lernen Und Evolution* (Springer-Verlag, Berlin Heidelberg New York, 1997). 3-540-62614-X
43. R. van Loon. *Machine learning explained: Understanding supervised, unsupervised, and reinforcement learning*. 2018. URL https://bigdata-madesimple.com/machine-learning-explained-understanding-supervised-unsupervised-and-reinforcement-learning
44. A. Mirfendreski, *Entwicklung Eines Echtzeitfähigen Motorströmungs- Und Stickoxidmodells Zur Kopplung an Einen HiL-Simulator* (Springer Fachmedien, Wiesbaden, 2017). 978-3-658-19328-7
45. W. Kinnebrock. *Neuronale Netze: Grundlagen, Anwendungen, Beispiele*. Oldenburg-Verlag, München, Wien, 2., verbesserte Aufl., 2018. ISBN 3-486-22947-8
46. J. von Neumann, First draft of a report on the EDVAC. IEEE Annals of the History of Computing **15**(4), 27–75 (1993). https://doi.org/10.1109/85.238389
47. D. Kriesel. *A brief introduction to neural networks*, 2007. URL http://www.dkriesel.com/en/science/neural_networks
48. K. Debes, A. Koenig, H. M. Gross. *Transfer Functions in Artificial Neural Networks: A Simulation-Based Tutorial*. brains, minds, media, 2005. URL http://www.brains-minds-media.org
49. S. Hochreiter, Y. Bengio, P. Frasconi, J. Schmidhuber. *Gradient flow in recurrent nets: the difficulty of learning long-term dependencies*. In S. C. Kremer, J. F. Kolen, editors, *A Field Guide to Dynamical Recurrent Neural Networks*. IEEE Press, 2001
50. A. Géron, *Hands-on Machine Learning With Scikit-Learn, Keras, And Tensorflow: Concepts, Tools* (And Techniques. O'Reilly Media, Sebastopol, 2019). 1492032646
51. G. Xavier, Y. Bengio. *Understanding the difficulty of training deep feedforward neural networks*. In Yee Whye Teh, Mike Titterington, editors, *Proceedings of the Thirteenth International Conference on Artificial Intelligence and Statistics*, volume 9 of *Proceedings of Machine Learning Research*, p. 249–256, Chia Laguna Resort, Sardinia, Italy, 2010. JMLR Workshop and Conference Proceedings. URL http://proceedings.mlr.press/v9/glorot10a.html
52. K. He, X. Zhang, S. Ren, J. Sun. *Delving Deep into Rectifiers: Surpassing Human-Level Performance on ImageNet Classification*. In *2015 IEEE International Conference on Computer Vision (ICCV)*, p. 1026–1034. IEEE, 2015. ISBN 978-1-4673-8391-2
53. R. Tadeusiewicz, *Sieci Neuronowe* (Problemy Współczesnej Nauki i Techniki. Informatyka. Akademicka Oficyna Wydawnicza RM, Warszawa, 1993). 838576903X
54. J.A. Freeman, D.M. Skapura, *Neural Networks: Algorithms, Applications, and Programming Techniques* (Addison-Wesley, Reading, Massachusetts, 1992). 0-201-51376-5
55. G. Breard, L. Hamel, Evaluating Self-Organizing Map Quality Measures as Convergence Criteria, in *ICDATA' 18*. ed. by R. Stahlbock, G.M. Weiss, M. Abou-Nasr (CSREA Press, United States, 2018). 1-60132-481-2
56. T. Kohonen, T. S. Huang, M. R. Schroeder. *Self-Organizing Maps*, volume v.30 of *Springer Series in Information Sciences Ser*. Springer-Verlag, Berlin, Heidelberg, 3rd ed., 2000. ISBN 978-3-642-56927-2

57. R.S. Sutton, F. Bach, A.G. Barto, *Reinforcement Learning: An Introduction*, 2nd edn. (Adaptive Computation and Machine Learning series. MIT Press Ltd, Massachusetts, 2018). 9780262039246
58. A. Tang et al., Canadian association of radiologists white paper on artificial intelligence in radiology. Canadian Association of Radiologists journal **69** (2), 120–135 (2018). https://doi.org/10.1016/j.carj.2018.02.002
59. P. Radhakrishnan. *A Medium publication sharing concepts, ideas, and codes: What are Hyperparameters? And how to tune the Hyperparameters in a Deep Neural Network?* Medium, 2017. URL https://towardsdatascience.com
60. R. Fullér. Fuzzy logic and neural nets in intelligent systems. In C. Carlsson, editor, *Information systems day*, TUCS general publications, p. 74–94, Turku, Finnland, 1999. Turku Centre for Computer Science. ISBN 951-29-1604-5
61. K. Wolff, R. Kraaijeveld, J. Hoppermanns. Objektivierung der fahrbarkeit. In K. Becker, editor, *Subjektive Fahreindrücke sichtbar machen, IV*, Haus der Technik - Fachbuchreihe. Expert, 2009. ISBN 978-3-8169-2936-9
62. T. Epelbaum. *Deep learning: Technical introduction*. URL http://arxiv.org/pdf/1709.01412v2
63. W. Zhu, Y. Ma, Y. Zhou, M. Benton, J. Romagnoli. *Deep Learning Based Soft Sensor and its Application on a Pyrolysis Reactor for Compositions Predictions of Gas Phase Components*. In M. R. Eden, M. G. Ierapetritou, G. P. Towler, editors, *13th International Symposium on Process Systems Engineering : PSE 2018*, volume 44, p. 2245–2250. Elsevier, 2018. ISBN 978-0-444-64241-7
64. A. Dertat. *Applied Deep Learning - Part 4: Convolutional Neural Networks*. Medium, 2017. URL https://towardsdatascience.com
65. D. Forsyth, J. Ponce, *Computer Vision: A Modern Approach* (N.J. and London, Prentice Hall, Upper Saddle River, 2003). 978-0-13-085198-7
66. J. Hui. *GAN - Some cool applications of GAN*. 2018a. URL https://jonathan-hui.medium.com
67. B. C. Kamble. *Speech Recognition Using Artificial Neural Network – A Review*, volume 3. 2016. 10.15242/IJCCIE.U0116002
68. M. A. Escabi. *Biologically Inspired Speech and Sound Recognition*, 2020. URL https://escabilab.uconn.edu/biologically-inspired-speech-and-sound-recognition/
69. J. Hui. *GAN — What is Generative Adversarial Networks GAN?* 2018b. URL https://jonathan-hui.medium.com
70. ActiveWizards Group LLC. *Comparison of top data science libraries for Python, R and Scala*. New York, USA, 2010-2020. URL https://activewizards.com/blog/comparison-of-top-data-science-libraries-for-python-r-and-scala-infographic/

MIX
Papier aus verantwortungsvollen Quellen
Paper from responsible sources
FSC® C105338

If you have any concerns about our products,
you can contact us on
ProductSafety@springernature.com

In case Publisher is established outside the EU,
the EU authorized representative is:
Springer Nature Customer Service Center GmbH
Europaplatz 3, 69115 Heidelberg, Germany

Printed by Libri Plureos GmbH
in Hamburg, Germany